Who's Asking?

Native Science, Western Science, and Science Education

Douglas L. Medin and Megan Bang

The MIT Press
Cambridge, Massachusetts
London, England

This book was set in Stone by the MIT Press. Printed and bound in the United States of America.

Library of Congress Cataloging-in-Publication Data
Medin, Douglas L.
Who's asking? : Native science, Western science, and science education / Douglas L. Medin and Megan Bang.
    pages   cm
Includes bibliographical references and index.
ISBN 978-0-262-02662-8 (hardcover : alkaline paper) 1. Indians—Science. 2. Indian philosophy. 3. Science—Philosophy. 4. Ethnoscience. 5. Science—Study and teaching. 6. Indians—Education. 7. Science—Social aspects 8. Science—Political aspects I. Bang, Megan, 1975– II. Title.
E59.S35M43 2013
303.48′3—dc23
2013018075

As a young person, I feel blessed, humbled, and honored to be working toward improving our communities in this capacity. I am thankful to all those I have walked with and learned from and especially grateful for my elders, dear friends, and family who have helped and supported me on this road, especially Antonia Wheeler Sheehy, Pam Silas, Doug Medin, and Carol Lee. To my parents, siblings, grandparents, aunties, uncles, nieces, and nephews, thank you for making me who I am, and shaping how I think and how I dream for our communities. To the Chicago Indian community, thank you for being the testament to resilience and spirit. And to my loves Lawrence, Nimkii, Miigis, and Akina: chi miigwetch.

—MB

President Obama calls Michelle Obama "his rock." My wife, Linda Powers, is my rock.

—DLM

Contents

Preface

We're going to start in the middle, and (to make things worse) we omit the beginning and the end. The beginning would take too long and the end is in the future, but this is a good point to pause and take stock.

The middle starts for Medin when Andrew Ortony suggested that Bang and Medin should meet. Bang was a Learning Sciences graduate student and (somehow) combined this with being director of education at the American Indian Center of Chicago. Medin was a professor of psychology and conducting research on mental models of nature on and around the Menominee reservation in Wisconsin. Bang was well acquainted with the Menominee Nation and Menominee people also make up an important part of Chicago Indian community.

Bang was interested in culture and learning and more specifically in science education and Medin in the role of culture and experience in children's understanding of the biological world. We talked, first about our separate interests and then about mutual interests. We ended up jointly teaching a course on research methods at NAES, Chicago campus (NAES = Native American Educational Services). There was also a NAES, Menominee campus, where Medin had taught, under the guidance of then-dean Karen Washinawatok. Eventually, we hit on the idea of a three-way research partnership involving NAES Chicago, NAES Menominee, and Northwestern University, and we were fortunate enough to receive a pilot grant from the Spencer Foundation to study culturally based science education (Karen Washinawatok was the NAES Menominee principal investigator).

This research partnership has survived some fairly substantial changes. For a variety of reasons, the Chicago component shifted to the American Indian Center of Chicago (AIC) and NAES, Menominee campus, became East–West University, Menominee Campus, before a further shift of the partnership to the Menominee Language and Culture Commission. The pilot grant and a Spencer Foundation Fellowship enabled Bang to complete her dissertation (Medin served on her committee) and she completed a postdoctoral fellowship at the Chèche Konnen Center at TERC in Boston, while also continuing to serve as the AIC principal investigator. Meanwhile, Washinawatok left East–West to become tribal chair for a year before shifting to her current position as director of the Menominee Language and Culture Commission. In part because of his emerging interest in science education, Medin shifted from 100 percent psychology to a split between psychology and the School of Education and Social Policy (thanks to the generosity of the latter). Part of this shift brought into contact different methodological traditions that have become generative for us—that is, a combination of design research (a staple methodology of the learning sciences), cognitive and developmental research methods, and field methods for studying cognition in everyday practices and artifacts.

Another shift was that we were not doing simply culturally based science education research but rather culturally and community-based research. This shift was inspired by scholarship focused on education and indigenous knowledge systems by Native scholars like Dr. Gregory Cajete (whose important book *Native Science* became something of a handbook) and Dr. Oscar Kawagley, and by the AIC and Menominee community members who were the backbone for our research. At least it was a shift for Medin— Bang's second-year project in graduate school proposed community-based design as a strategy for design research that has been generative in the work we have done and currently are doing. This research has been supported by the National Science Foundation and by the Christian A. Johnson Endeavor Foundation.

As our projects continued to unfold, we increasingly saw the potential of combining developmental studies assessing biological cognition with observations on informal (out-of-school) science learning and design experiments aimed at putting what we were learning into practice and then reflecting on and evaluating these practices in the service of improved design. It was also evident that community- and culturally based science is not simply about knowledge and beliefs, but also is reflected in the most basic of science-related practices. We believe that Native science is not a historical science displaced by "modern" science, but rather a distinct perspective on doing science that is more relevant today than ever. More broadly we believe that Western science has been, and is, seriously limited by the lack of diverse perspectives of practitioners of science and science education. As humankind continues to struggle toward a viable future for life on Earth, we have come to see these issues for science and its practitioners as among the most critical we face.

We are deeply indebted to many, many people for this project. Karen Washinawatok (Menominee) has been an advisor, partner, friend, and mentor at every step. Sandra Waxman has been a co-principal investigator and the brains behind many of our developmental studies, as well as a wonderful colleague and friend. Donna Powless* (Oneida) and Shannon Chapman (Menominee) welcomed us to the Menominee Tribal School, and Shannon was a principal investigator on one of our grants before becoming principal of the school. Pamala Silas (Menominee/Oneida), former CEO of the American Indian Science and Engineering Society and leader in the Chicago community, has been and is a fountain of wisdom and inspiration. Joe Podlasek (of LCO Ojibwe and Polish descent), president of the National Urban Indian Family Coalition and former executive director of the American Indian Center, has supported the work in the Chicago community. Cynthia Soto (Lakota/Puerto Rican) and Ananda Marin (Choctaw, African American, European American descent) have demonstrated ongoing leadership and kept us grounded throughout the course of our work. Most importantly, we both have benefited from the generosity and advice of elders, including more than a few who have passed on.

We acknowledge a debt to the community members at the American Indian Center of Chicago and on the Menominee reservation, including the members of the Menominee Language and Culture Commission. Any listing would be incomplete but we want to at the very least mention the following people who have either held important leadership roles or have deeply impacted the direction of our work: From the AIC: Jasmine Alfonso (Oneida/Menominee), Alexis Bellinger* (Ojibwe), Corey Brown, Julia Brownwolf * (Lakota), Jannan Cotto (Odawa/Ojibwe), Lawrence Curley (Ojibwe/Navajo), Alma Enriquez (Sioux/Chickasaw), Lori Faber (Oneida Tribe of Wisconsin, enrolled member), Gennafer Garvin (Ho-Chunk/Meskwaki), Tandy A. Garvin (Ho-Chunk Nation), Thomas Heaton (Lakota), Adam Kessel (Lakota/Italian descent), Mike Marin* (Navajo, Pueblo, Washo), Jennifer Michals (Potowatomi/Ojibwe), Mavis Neconish (Omaeqnomeniahkiw), Ashlee Pinto (Tunica-Biloxi), Heather Reed (Menominee), Raven Roberts*

(Potowatomi/Mi'kmaq/Ojibwe), Anthony Roy* (M'chigeeng First Nation), Skip and Babette Sandman* (Ojibwe), Angel Starr (Omaha/Odawa/Arikara), George Strack (Miami tribe of Oklahoma), Eli Suzukovich (Little Shell Band of Chippewa-Cree and Serbian), Deborah Valentino* (Oneida/Menominee), Sally Wagoner and Antonia Wheeler-Sheehy* (Blackfeet), and Negwes White* (Ojibwe, Navajo). From the Menominee Nation (Menominee unless otherwise noted): Amy Almadinger* (Ojibwe), Richard Annimetta, Liz Arnold, Chris Caldwell, Michael Chapman, Doug Cox, Carol Dodge, Robert Fernandez, Tony Fernandez, Julie Kaquatosh (Ho-Chunk), Tammy Lyons, Kateri Merino, Amy Miller-Cox, Connie Rasmussen (Oneida), Brett Reiter, Julie Schlichting, Glenda Tahmahkera, Mike Waukau, Rose Wayka, Stuart White (Menominee descent), Allen Washinawatok, Dot Wescott, Melissa Wescott, Sara Wescott. From Shawano Wisconsin: Tina Burr and Kay Fredrick.

This monograph was improved by the comments and suggestions of Sonya Sachdeva, Rumen Iliev, bethany ojalehto, Jennie Woodring, Brock Ferguson, Barry Schwartz, and several anonymous reviewers. Jennie also was the key person for manuscript preparation and editing help. Most of all we thank her for not complaining about our unreasonable requests.

Both authors are grateful for the support in producing this work of the National Science Foundation under Grant Numbers DRL0815020, DRL0815222, DRL1109210, and DRL1114530, as well as the Christian A. Johnson Endeavor Foundation (award letter 3-13-06) and the Spencer Foundation (MG200300137). Medin also thanks the National Science Foundation for support under Grant Number SES0962185 and the Cattell Foundation for a sabbatical award to work on this monograph.

1 Introduction: Who's Asking?

"Researcher 'diversity'—this is a showstopper for me."

That opening comment needs a bit of explanation. It comes from a friend and outstanding scholar who has done pathbreaking cultural research. The context was a team of us writing a short essay to the National Science Foundation (hereafter NSF) on why the behavioral and social sciences should have a strategic priority of increasing the range of their study populations.

Why this priority? Because broad claims about human psychology and behavior based on very narrow samples from Western societies are regularly published in leading journals. An analysis of the top journals in six subdisciplines of psychology from 2003 to 2007 (Arnett 2009) revealed that 96 percent of participants were from Western industrialized countries. These samples also reflect the country of origin of the authors—99 percent of first authors were at universities in Western countries. Furthermore, studies with adults are largely studies of college students attending these same universities and studies with children are largely drawn from the middle-class communities that are close by.

Of course, this narrowness of samples may not be a problem if people are pretty much the same the world over. But cross-cultural studies show that claims about basic cognitive processes derived from studies with undergraduates often do not generalize well to other populations (Atran and Medin 2008). In fact, a recent extensive review of cultural comparisons involving a wide range of phenomena including visual perception, judgments of fairness, categorization, spatial cognition, memory, moral reasoning, and self-concepts (Henrich, Heine, and Norenzayan 2010) strongly suggests that these "standard" research participants actually may be unusual or "outliers" compared with the rest of humankind.

That was half of our argument to NSF. The other half was about the diversity of scientists carrying out the research. We argued that diversity was a good thing, not solely on grounds of fairness and equity, but also because science would be the better for it. That was the showstopper for the person quoted above.

And he wasn't the only one. Although support for study population diversity was essentially universal among the researchers we approached (this may reflect our choice of who to approach), researcher diversity was much less enthusiastically received. One fellow researcher commented, "I would not add diversity of experimenters in here, because I think it is distracting," and another suggested that "the diversity requirements currently in place have turned into a burden on investigators with no positive impact on anything, so I am not keen on more of those efforts to increase diversity."

Here's another related story. During the year when Carol Lee was president of AERA (American Educational Research Association), Medin and Lee approached the editor of *Science* magazine, proposing that we contribute an editorial on the need for diversity in science. Our proposal was initially accepted and we were given a target date for submission. Our thesis was, as previously noted, that diverse science makes for better science. After we sent in our editorial we did not hear from *Science* for months and months. Finally we inquired and were told that the feeling was that the topic of diversity was a bit "shopworn and stale" and that perhaps they would get back to us sometime in the future. Again, apparently a showstopper. Of course there's also the possibility that what we had written just wasn't very good.

Is there any connection between researcher diversity and the effectiveness of scientific research? Our friends certainly are divided. The scientist for whom diversity was a showstopper went on to elaborate his argument: "Economics is filled with Indians, Europeans, Latin Americans, and East Asians. It's much more diverse than psychology. The most important young economist in the world today is Turkish (MIT). Is economics better theoretically for it? (If your answer is 'yes,' how?) Anthropology did affirmative action for PhDs a long time ago. Grad students are recruited from everywhere. Mostly, they apply postmodernist critiques to their hometowns. Lots of diversity, no scientific progress."

The preceding observation may be accurate if the students who succeed in the sciences are only those who buy into the majority, Western orientation toward their discipline.[1] But as we'll see later on, there are important exceptions to this notion.

Elissa Newport, a close friend (of Medin's) and a psycholinguist, had a very different take on things. She wrote, "I didn't answer your previous email earlier only because I was trying to figure out if there was an easy way to broaden the letter a bit, to include a couple of diversity issues close to my own heart—including sign languages and deaf communities as equally important for study as spoken languages (at the moment everyone studies spoken languages, and then the specific investigators who work on sign languages do that as a special topic), and also as emphasizing training native members of other languages and cultures so that they can be the ones who do research on their own culture or be the ones who collaborate on such studies (in the sign language field, most labs have some underling who is deaf and employed as a lab tech, but very few researchers [like Ted] are actually trained and working in the field)." (Ted is a deaf psycholinguist and Elissa's husband.)

Aside from supporting diversity as a matter of equity, does scientists' diversity really matter? We think that it does. It matters because it touches on key issues of the relation between science and society, between science as authority and public policy, and, as we will see, because it raises foundational issues for science education, especially for students of color. These claims are not obvious, and it will take some time for us to develop our arguments and to support them.

To examine the relationship between researcher diversity and science we have to examine the nature of science more closely. Although it is a bit of a caricature, we will contrast two competing narratives about science. In one corner we have the view that science is objective, value-neutral, and acultural. Although individual scientists may have biases, the sociology of science and the associated competition of ideas lead science eventually to truth.

Furthermore, it doesn't matter how a scientist would like the world to be; what counts is how it is, and that's what science aims to discover. This grounding in the world ensures the unity of science. Although there are always scientific disputes, scientists believe that there is one correct account of how things are in the world (and what makes it correct is that it *is* how things are). In short, (the nature of) science is settled.

Some people in this corner also are fond of the story of the development of science according to which it basically started in Greece, was nurtured in Europe during "the Enlightenment" and the associated triumph of reason, and eventually grew into modern science (a.k.a. "Western science"). Only in the West has science been cleanly severed from the irrationality of magic and superstition. That's the gist (a.k.a. stereotype) of the first narrative.

In the other corner are scholars who question all of these claims. There's more variability of opinion in this corner, but membership tends to include the following other claims. Scientists do not shed their cultures at the door and their practices reflect their culture's values, belief systems, and worldviews. Unity of science is not assumed—scientific models, theories, and representations highlight some aspects of reality but may ignore or conceal others. On this pluralistic view, different theories of the same phenomena may be useful for different purposes (like wave versus particle theories of light). In this view, each theory is only a slice or perspective, and different aspects of reality may be relevant to them.

Some in this corner see a science dominated by white males as limited in its perspective, and consequently they are supportive of researcher diversity. (A few see white male science as biased and incorrect, but the majority just sees it as [potentially] different and limited.) Finally, the history of science they tell is not nearly so Eurocentric and triumphalist in its tone.

Many contemporary scientists find themselves in conflict. The unity of science / one truth / no bias position sounds unrealistic, even as an ideal (can there be a science that expresses or reflects no values?). But the other position seems to go too far—can it rule out creation science and other pseudosciences as meeting the standards for science? So scientists may consider themselves to be facing a choice between an unrealistic ideal and something of an "anything goes" Pandora's box.

Our opinion is that contemporary science and science education are being held back by this pattern of dichotomous thinking and defensiveness. Many potential scientists are turned off by the prospects of a science without values and approaching nature from a remote, dispassionate perspective. This sort of science is sometimes not obviously relevant to their daily lives or fails to satisfactorily meet pressing needs. These feelings may be especially salient for minority scholars who feel a responsibility to "give back" to their communities.[2] The same can be said for science education, which also treats science as settled. But maybe we're getting ahead of ourselves.

What This Book Is About

So far we've talked about the culture of science (or its presumed cultural neutrality), but equal focus is on the science of culture and its implications for science education. We see the culture of science and the science of culture as closely intertwined, as two sides of the same coin. Recent National Research Council reports have emphasized the idea that formal science education will be more successful if it can take advantage of the science learning that takes place in informal contexts, outside of school (Bell et al. 2009). There is substantial evidence, some of which we will describe in detail later on, supporting the idea that children come to school with knowledge, orientations, values, and practices that are relevant to science learning and that reflect their own culture. When these orientations are supported, students are more engaged, identify with, and are more successful with science than when these orientations are ignored or discouraged (Bell et al. 2009; Bransford and Brown 2000).

How are things working in the United States at the moment? The answer is not very well for anyone, and especially not very well for U.S.-born students of color. Although the demand for science and engineering workers has been growing very rapidly, the supply has been extremely limited. The growth in science and engineering doctorates awarded in the U.S. over the past decade has almost exclusively been driven by increases in foreign students. As noted in the National Academy of Sciences report *Rising above the Gathering Storm* (National Academy of Sciences et al. 2007), in the year 2000 the United States ranked twentieth out of twenty-four countries in the percentage of 24-year-olds who had earned a first degree in the natural sciences or engineering. That report recommended setting the goal of increasing the percentage of 24-year-olds with these degrees from 6 percent to at least 10 percent, a standard already met or exceeded by many of the other twenty-three countries. The United States will be less competitive globally if fails to develop or otherwise limits the pool of S&E (science and engineering) scholars.

Let's introduce a demographic twist on this goal. People of color represent a significant segment of the US population (about a third), and this segment is projected to grow substantially, such that the United States will likely have a majority-minority population by mid-century. If S&E degrees are going to grow from 6 percent to 10 percent, then much of that growth will need to come from minority scholars. In the words of the 2010 NAS report *Expanding Underrepresented Minority Participation*, "This growth rate provides an opportunity as well as an obligation to draw on new sources of talent to make the S&E workforce as robust and dynamic as possible."

This same report lays out the current sobering picture with respect to underrepresented minority participation in S&E ("underrepresented minority" does not include Asian Americans, who are the most overrepresented group in S&E). To quote from that report:

But we start from a challenging position: underrepresented minority groups comprised 28.5 percent of our national population in 2006, yet just 9.1 percent of college-educated Americans in science and engineering occupations (academic and nonacademic), suggesting the proportion of underrepresented minorities in S&E would need to triple to match their share of the overall US population.

Underrepresentation of this magnitude in the S&E workforce stems from the underproduction of minorities in S&E at every level of postsecondary education, with a progressive loss of representation as we proceed up the academic ladder. In 2007, underrepresented minorities comprised 38.8 percent of K–12 public enrollment, 33.2 percent of the US college age population, 26.2 percent of undergraduate enrollment, and 17.7 percent of those earning science and engineering bachelor's degrees. In graduate school, underrepresented minorities comprise 17.7 percent of overall enrollment, but are awarded just 14.6 percent of S&E master's and a miniscule 5.4 percent of S&E doctorates. Only 26 percent of African Americans, 18 percent of American Indians, and 16 percent of Hispanics in the 25–29-year-old cohort had attained at least an associate degree.

But again, the statistics are even more alarming for underrepresented minorities. These students would need to triple, quadruple, or even quintuple their proportions with a first university degree in these fields in order to achieve this 10 percent goal: at present, just 2.7 percent of African Americans, 3.3 percent of Native Americans and Alaska Natives, and 2.2 percent of Hispanics and Latinos who are 24 years old have earned a first university degree in the natural sciences or engineering.

Recent data from the Higher Education Research Institute (HERI) at UCLA shows that underrepresented minorities aspire to major in STEM [science, technology, engineering, mathematics] in college at the same rates as their white and Asian American peers, and have done so since the late 1980s. Yet, these underrepresented minorities have lower four- and five-year completion rates relative to those of whites and Asian Americans. That a similar picture was previously seen in data in the mid-1990s signals that while we have been aware of these problems for some time, we, as a nation, have made little collective progress in addressing them. (National Academy of Sciences et al. 2010, 3–5)

As the report goes on to note, this gloomy picture is still worse if we consider who is able to obtain research grants from agencies like the National

Institutes of Health (NIH) and the National Science Foundation. The figure of 6.8 percent of doctoral scientists who are minorities (again, non-Asian) shrinks to about 3 percent when it comes to NIH and NSF funding for research.

The Specific Case of Native Americans

Since Native Americans will be the main minority group that we will focus on in this monograph, we will take the time and space to describe their corresponding statistics on underrepresentation in further detail. For Native Americans the picture we have presented remains accurate or is even more extreme. There is chronic underrepresentation of indigenous people in science-related fields. Native Americans drop out of high school at the highest rates of all ethnic groups in the continental United States, and only about 6 percent receive bachelor's degrees (National Academy of Sciences et al. 2007; Pavel and National Center for Education Statistics 1998).

Nowhere is the problem more apparent than in science learning. For example, between 1997 and 2007, Native people represented an average of 0.63 percent of the total number of bachelor's degrees and an average of 0.48 percent of the doctorates awarded in science and engineering (National Science Foundation 2007). The 2010 census found that about 1.7 percent of the U.S. population identified themselves as American Indian or Alaskan Native. Thus, these figures indicate that Native people are substantially underrepresented at the college level and even more underrepresented at the doctoral level. Even without taking into account the age distribution of Native Americans, the number of doctoral degrees would need to more than triple to be proportionate. Considering that the Native American population has many young people (especially compared with the aging U.S. European American population), a more accurate figure might be four or five times the present rate.

These figures are relevant to the broader picture of minorities in STEM (science, technology, engineering, mathematics) disciplines, but they are especially relevant to Native nations who are confronted with environmental and resource management issues and sometimes are forced to rely on outside expertise from people who may not share their values or have the cultural and historical understandings needed to provide effective advice. These shortages are equally relevant to urban Indian communities as they also face issues where STEM expertise is needed, not only in the health and medical sciences but also in environmental policy debates.

The shortage of degreed expertise within Native communities contributes to and perpetuates struggles with education and educational achievement, adequate economic development, the enhancement of community health, and community-based governance of resource management. To improve the circumstances that affect indigenous communities in ways that are likely to have a sustained impact requires that we improve the educational experience and attainment of Native peoples, especially within STEM education.

Although there are important general questions concerning how to foster educational achievement, there is reason to think that some of the difficulties are specific to science education. Here is an example from the Menominee reservation in Wisconsin where we have conducted research for the last fifteen years. A 2005 report of statewide, standardized tests indicated that in fourth grade Menominee children scored above the national average in science and it was their best subject; by eighth grade, however, Menominee children scored below the national average in science and it was their worst subject (see Bang, Medin, and Atran 2007; Grigg, Lauko, and Brockway 2006). These striking findings could be interpreted in a number of ways, but we think they indicate that the orientations toward the natural world that Menominee children bring to the classroom are not clearly reflected and supported in science teaching.

In our view problems with achievement are more complicated than simply knowing or not knowing "science content" (see Demmert and Towner 2003 for a review of studies with Native American populations). Instead, the key issue might be "who owns science?" If science is seen as or implicitly operates as a franchise owned and operated by white people, then minority students may well not identify with science or see it as relevant to their lives.

So let's sum up for a minute. There is something about the educational experience of minorities that leads to a situation where minorities are severely underrepresented in STEM fields. This does not appear to arise from differences either in initial abilities or initial interest. We think that one factor in this underrepresentation derives from a clash (where the clash takes multiple forms) between the cultural values, orientations, and practices of minority students in general, and Native American students in particular, and the cultural values, practices, and orientations currently privileged in science education. This is why we believe that it is critical to understand the cultural orientations and practices that children bring to the classroom, and why we need a more robust science of culture, especially with respect to cultural orientations toward the natural world. As we mentioned before, that's one side of the coin.

On the other side of the coin, participation in science is limited by the common presumption that science is intrinsically acultural and value-free. This stance could help explain why students in general seemed to be turned off to science and engineering (see Aikenhead 2006). In Medin's decades of contact with Northwestern University undergraduates he has often encountered students who are planning to go to medical school, and they often have a story to tell linking their commitment to some situation involving a relative or close friend confronted with a medical condition. Medin has *never* had an undergraduate majoring in a STEM discipline tie their commitment to any real-world problem, though the recent upsurge of interest in environmental sciences may change that.

We believe, however, that this side of the coin is actually a science reflecting the cultural values of its practitioners, not some "neutral" science. From our perspective it is critical to examine each side with respect to the other—we believe that we cannot deeply improve science learning and identification with science (side 1) without querying science and science education as a set of cultural practices (side 2).

With respect to science learning there are once again two competing narratives. One sees science as some objective thing, and assumes either implicitly or explicitly that cultural/ethnic differences in learning and taking up science have nothing to do with the science side and everything to do with the cultural side. According to this perspective, one might do culturally based research from the point of view of identifying barriers to (minority) participation in science, such as unequal access to resources or lack of role models.

An example of this orientation may be the NAS report on increasing minority representation in science and engineering that we cited earlier. It made the following recommendations:

(1) prepare America's children for school through preschool and early education programs that develop reading readiness, provide early mathematics skills, and introduce concepts of creativity and discovery; (2) increase America's talent pool by vastly

improving K–12 mathematics and science education for underrepresented minorities; (3) improve K–12 mathematics and science education for underrepresented minorities overall by improving the preparedness of those who teach them those subjects; (4) improve access to all postsecondary education and technical training and increase underrepresented minority student awareness of and motivation for STEM education and careers through improved information, counseling, and outreach;[3] (5) develop America's advanced STEM workforce by providing adequate financial support to underrepresented minority students in undergraduate and graduate STEM education; and (6) take coordinated action to transform the nation's higher education institutions to increase inclusion of and college completion and success in STEM education for underrepresented minorities. (National Academy of Sciences et al. 2010, 171–181)

These laudable recommendations illustrate the point that cultural differences in participation in science do not necessarily lead to a deficit model (as in "What's wrong with these people that makes them unable or unwilling to become scientists?"). Note also, however, that the NAS report does not query science itself, or the practices associated with it.

Here's the competing position. The practices associated with science and science education reflect the cultural values and orientations of the practitioners. Thus the answers to scientific questions depend on who's asking, because the questions asked and the answers sought depend on who's asking, even when all parties adhere to rigorous research methods. We are not claiming that science is subjective, but rather that scientific practices embody values and perspectives, and these values and perspectives may vary across factors like gender, social class, and culture.

Participation and achievement in science are mediated by a complex set of sociocultural factors not often recognized in such science equity efforts. A certain claim is that one's social world and context shape values, skill sets, and expectations (Nasir and Hand 2006). Thus, the act of exposing all individuals to the same learning environments does not result in science equity, because the environments themselves are designed in a manner that supports the cultural repertoire of the dominant culture. (We'll have quite a bit more to say about this in later chapters.) Because of this neglect of cultural issues, science instruction may often privilege and support the science-related practices of middle-class European Americans and frequently fails to recognize the science-related practices associated with other groups (Bang et al. 2012; Lee and Fradd 1998; Lemke 1990; Moje et al. 2001; Nasir and Hand 2006; Warren et al. 2001). In other words, traditional approaches to science education reflect cultural assumptions and framework theories that may be alien to students and scholars of color.

Ballenger and Rosebery (2003) note that educators often hold stereotyped notions of what counts as scientific reasoning and privilege a subset of sense-making practices at the expense of others. For example, scientists regularly use visual and discursive resources whereby they place themselves in physical events and processes to explore the ways in which they may behave (Ochs, Gonzales, and Jacoby 1996; Wolpert et al. 1997). Yet these same practices often are not recognized as useful or a part of science in the classroom, and this lack of recognition has the effect of marginalizing students' home discourses (Rosebery and Hudicourt-Barnes 2006). These and other findings undermine the view that professional scientific practices are largely disassociated from forms of experience and practice in the everyday world (Warren et al. 2001).

This competing narrative requires that researchers examine the *relationships between* cultural practices and values that children bring to the classroom and the cultural practices associated with science and science education. A more conservative position might be to argue that the key is not science per se but rather science education practices. The argument would be that science instruction emphasizes those practices that are common in white middle-class communities, if only because the teachers themselves typically are white and middle-class. One example of these practices, not specific to science learning, is the IRE (inquiry-response-evaluation) form of interaction (What do we breathe out? Carbon dioxide? Right!). IRE is commonly used in classrooms, and anyone who has spent time in middle-class communities will recognize that it is also common outside of school. Hence it is a form that middle-class children are familiar with before they enter school. In other cultural communities it may be strange to have knowledgeable adults ask children questions that the adults already know the answers to. Although we agree that science education practices are critically important, we reject this conservative view in favor of the claim that we must query science itself.

Overview

Our plan for this monograph is as follows. Our overall thesis is that scientist diversity provides new perspectives and leads to more effective science. We will briefly review the history of science, primarily to critique the view that science had its unique origins in the West. As part of the analysis of the cultural character of science practices we will also take up some arguments and evidence from feminist science (we could and perhaps should have been mentioning gender all along, as science often appears to be the province of men) and from studies of the sociology of science from a cross-cultural perspective. Furthermore, we will suggest that the sciences may well need a pluralistic account and that there are compelling reasons to doubt the unity of science.

The heart of our monograph is a case study of Native American and European American orientations toward the natural world. This is intended both as a concrete example of how culture affects science-related practices and as an application to science education. We will argue that Native Americans and European Americans manifest distinct ways of situating themselves with respect to the rest of nature, ways that are both explicitly articulated and implicit in practices. These differences are consistent across a wide range of converging measures and evident in children as well as adults.

In a nutshell, the European American model sees humans as separated from nature and the Native American model sees humans as a part of and living in relationships with the rest of nature. We will argue and provide evidence that these differences in epistemologies have wide-ranging implications for science-related values and practices.

Then we turn to our efforts to develop culturally and community-based science education programs, in both a rural and an urban context (on the Menominee reservation in Wisconsin and at the American Indian Center of Chicago). These efforts provide something of a testimonial for our arguments. Along the way we will necessarily take up the history of Indian education in the United States and a series of methodological and ethical issues associated with research in tribal contexts. Although our story is complicated and the forms of evidence wide-ranging, we believe that the overall picture is clear and compelling in its account of the culture of science and the science of culture.

To make our story a bit more clear and coherent we'd better say something about who we are. Dr. Megan Bang is Professor of Learning Sciences and Human Development at the University of Washington. On her mother's side she is Ojibwe and on her father's side Italian. Until recently, she was a longtime member of the Chicago Indian community. Formerly she was Director of Education at the American Indian Center of Chicago (AIC) for more than a decade. At the AIC she co-founded a very effective tutor/mentor program, named Positive Paths. These professional experiences motivated her to return to graduate school where she met Medin. She earned her PhD in learning sciences at Northwestern University. In part, the inclusion of the urban intertribal Chicago community in our research evolved because Bang attended graduate school and Medin served as one of her advisors.

Although Bang had personal relationships with various members of the Menominee Nation prior to graduate school, as a graduate student she became professionally involved with the Menominee Nation through Medin and has now been working with the Menominee community for the past nine years. This has been coupled with continuous involvement with the AIC community and Bang has been the AIC principal investigator on several grants. Bang has taught at Northwestern University and Bang and Medin have co-taught a course at NAES (Native American Educational Services), Chicago campus. Bang has been a respected leader in the Chicago Indian community and, in Medin's view, has a special gift for empowering others.

Douglas Medin is Professor of Psychology and of Education and Social Policy at Northwestern University. On his father's side he is Swedish and German and on his mother's side Scottish and French. He is fond of saying that he has gone from being a first-rate cognitive psychologist to being a second-rate cultural anthropologist, before assuming his present role as a third-rate educational researcher. The subject matter correlates of these shifts go from the psychology of concepts, to studies of culture and cognition, to efforts to engage community-based science education both at the AIC and on the Menominee reservation in Wisconsin.

Medin has formed many relationships with Menominee elders, adults, and youth over these years, participated in many community events, and has supported and worked on community issues outside of research projects. Since these projects began Medin has established a similar relationship with the Chicago intertribal community and until recently served on the AIC board of directors. He has taught courses at NAES, Chicago campus; NAES, Menominee campus; and the College of the Menominee Nation.

Andrew Ortony, professor in the Northwestern School of Education and Social Policy, introduced Bang and Medin when Bang was a learning science graduate student. We soon found that we were kindred spirits and started working together at NAES, Chicago campus, the Menominee reservation, and then the AIC. Indeed we've worked together for more than a decade and been co-principal investigators on a fair number of grants. This monograph is one outcome of our collaboration. Working together closely and well makes relative contributions invisible and each of us has great respect for the other. We are also respectful and humbled by the individuals and communities that have helped us every step of the way.

2 Unsettling Science

If we were to picture physical reality as a large blackboard, and the branches and shoots of the knowledge tree as markings in white chalk on this blackboard, it becomes clear that the yet unmarked and unexplored parts occupy a considerably greater space than that covered by the chalk tracks. The socially structured knowledge tree has thus explored only certain partial aspects of physical reality, explorations that correspond to the particular historical unfoldings of the civilization within which the knowledge tree emerged. Thus entirely different knowledge systems corresponding to different historical unfoldings in different civilizational settings become possible. This raises the possibility that in different historical situations and contexts sciences very different from the European tradition could emerge. Thus an entirely new set of "universal" but socially determined natural science laws are possible. (Harding 1994, 320)

If you're like us, you've been exposed to a series of related stories about the history of science, ending up with "modern science." The tenor of these tellings often is very much of the form, "people used to think x, but now we know y." Although there are new scientific discoveries every day, science itself is now treated as "settled." The history of science is, in part, the story of its struggles with other ways of knowing, such as religion. And it was not until the scientific revolution, roughly between 1500 and 1700, featuring luminaries such as Copernicus, Galileo, Kepler, Descartes, and Newton, that science and its methods became firmly established as a distinct, objective way of knowing. According to this perspective, the history of modern science is more or less equivalent to the history of Western science.

In sharp contrast to this view, there are a number of scholars who have argued that the history of science as typically taught in schools is little more than Eurocentric ethnoscience. This consensus reveals the story of science to be "unsettled." In this chapter we will review what we take to be the "standard view" of science that sees Western science as triumphant. Then we will turn to the scholarly critique of this view and begin to explore its implications.

The (Cultural) History of Science (from a Western Perspective)

The Beginnings

Here's our stereotyped version of the history of science: the roots of modern science began with ancient Greece. Aristotle made systematic observations of the natural world, and his works include the earliest known formal study of logic. Plato, along with his teacher Socrates and his student Aristotle, helped lay the foundations of Western science (and philosophy). Later on, Greeks like Galen and Hippocrates laid the groundwork for modern medicine.

As the story goes, following the decline of the Roman empire around 400 CE until about 1400 CE (1000 by some modern accounts), there was a period of cultural stagnation and deterioration in Europe where very little occurred relevant to science or other forms of knowledge. Arab scholars and Catholic monks are given some credit for the preservation of works from ancient Greece, but all in all, the dark ages were pretty dark.

But then came the Renaissance or rebirth of European intellectual life. Scholars returned to and translated classic works written in Greek and Latin. Johannes Gutenberg invented the printing press around 1450 and for the first time texts could be widely available. Some historians suggest that the "Black Death" (the plague) that decimated fourteenth-century Europe led people to focus more on their lives on earth rather than the afterlife, an orientation that encouraged humanism. Stars of the Renaissance include Michelangelo and Leonardo da Vinci. Da Vinci was both an artist and a scientist, and some credit him with the development of the scientific method and a focus on empirical evidence (others credit Francis Bacon). This period was also associated with the development of capitalism, exploration, and the beginnings of European colonialism.

The Renaissance set the stage for the "Enlightenment" in the seventeenth and eighteenth centuries. The Enlightenment embodied freedom of inquiry and a willingness to challenge tradition in the search for truth. Tradition included superstition (witchcraft trials were common in this period) and religion. Science dared to question both. Every American child learns in school about how the Catholic Church punished Galileo for suggesting, as had Copernicus, that the Earth is not the center of the universe. The ultimate triumph of science was also the triumph of rationality, and some call this period the Age of Reason.

The Age of Reason was associated with huge advances in science, technology, mathematics, and engineering. These advances supported and enabled colonialism and the associated spread of ideas. Many readers may be familiar with Jared Diamond's 1997 book *Guns, Germs, and Steel*, where he takes up the task of explaining why Eurasian civilizations have survived and conquered others rather than the reverse. Two of his big three, guns and steel, are forms of technology, and colonialism was significantly fostered by scientific developments in the West. Indeed, one could frame colonialism as the bringing of "better ways of living" to less advanced peoples (we hope the irony in this description is obvious).

Clearly we are guilty of caricature. We also learned that European explorers brought back corn, tobacco, and potatoes from the New World. These crops had been cultivated by indigenous Americans, so New World peoples had obviously employed empirical methods for learning about the world. Indeed, even today indigenous insights into medicinal uses of plants (e.g., salicylic acid from willow as a precursor to aspirin) and related "traditional ecological knowledge" are widely acknowledged as important, though some scholars withhold their approval until results can be "verified by modern science." More broadly, the argument is not that other cultures did not engage in scientific (empirical) practices but rather that only modern (Western) science succeeded in divesting superstition from science.

The Science of Recent History

Witchcraft trials are no longer common and our lives are enormously dependent on technological advances. We look to science for the solution to many problems, and science supports better ways of living in many spheres of life. Western science is just science. The United States has supremacy in science (currently anyway), in part because our freedoms include a freedom of inquiry.

Science is also a quest for truth. It is acultural, unbiased, and value-neutral. It tells us the way things are and perhaps how they could be, but science does not tell us how things should be. Science can inform policy but itself reflects no policy. When Abraham Lincoln established the National

Academy of Sciences in 1863 as advisors to the nation, the act of incorporation stipulated that "the actual expense of such investigations, examinations, experiments, and reports to be paid from appropriations made for the purpose but the Academy shall receive no compensation whatever for any services to the Government of the United States." This stipulation underlines the neutrality of science. Science is public, self-correcting, and relies on the scientific method, not opinion. It is constrained only by the nature of reality.

That's more or less what we were taught in school. There are four components of this history that will be central to our coming analysis. First, there is the concrete historical question of who did what when. We were also taught in school that Columbus discovered America, but both logic and history reveal that this cannot be literally true.

The second component of our analysis concerns rationality and the relation between science and other ways of knowing. This component is closely related to the third issue, the historical development of the methods of science. Our short, stereotypic history suggests that only Europeans were able to escape the "irrationality" of superstition and the associated projection of cultural values onto nature in order to see the world the way it objectively is. The fourth and perhaps most central component concerns the purity and objectivity of science. Is the history of science the history of the triumph of human intelligence over ignorance? Is it humankind's greatest success story?

To set the stage for what follows we summarize with a distillation of ideas about science provided by Traweek (1996). She says:

I want to recapitulate, perhaps too tersely, a few of those older notions we usually first encountered in the pages of high school and undergraduate textbooks, or perhaps in museums and on television, that often survive in our minds. [Here we abbreviate her list a bit.]

* Galileo's ideas were rejected by the Vatican because they challenged Catholic religious beliefs of the time.
* Francis Bacon developed the idea of a laboratory and codified the procedure for research now called the scientific method.
* The printing press made possible the accurate reproduction and circulation of experimental data.
* Isaac Newton invented the idea and the means of using mathematics to analyze experimental data.

* Scientific method is based upon skepticism.
* Scientific method identifies and controls all variables in an experiment.
* Scientific knowledge is amassed progressively and cumulatively.
* Scientific theories and data are rejected when subsequent efforts at replication fail.
* New scientific theories are accepted because they explain more experimental data more economically than their predecessors.
* Scientific reasoning proceeds by deduction and induction; hypotheses are deduced from existing experimental data and experimental data are tested against hypotheses inductively.
* Scientific research is made objective by eliminating all biases and emotions of the researchers.
* Scientific research is neutral with respect to social, political, economic, ethical, and emotional concerns.
* Scientific research has an internal intellectual logic; there is an external social, political, economic, and cultural context for science that can only affect which scientific ideas are funded or applied.
* Improvements in the quality of human life and the duration of human life during the past two hundred years are due primarily to the application of scientific discoveries.
* Basic research and applied research are easily differentiated.

As we will see, there are reasons to question most if not all of these assumptions and, more broadly, the associated expressed and implied Eurocentric narrative.

Counterpoint: Western Science versus Western Ethnoscience

Scientific Innovations

Have you ever read a news article describing Russian claims that important discoveries and inventions were first made in the USSR rather than the West? We have, and we admit to being amused by their audacity and obvious inaccuracy. However, that initial reaction reveals our enculturation into a particular perspective. The main problem with what we were taught about the history of science in school is that much of it isn't true.

First consider the issue of precedence. As Harding (1994, 308) notes:

What has been ascribed to the European tradition has been shown on closer examination to have been done elsewhere by others earlier. Although Harvey is credited with discovering the circulation of blood, Ibn al-Nafis studied the human body and beat William Harvey by three and a half centuries. Ancient Iraqi schools taught algebra and geometry, knew what is now called the Pythagorum theorem as early as

1700 BCE, and knew the value of *pi* (Hobson 2004). All six of the classical trigono-
metric functions had been defined and tabulated by Muslim mathematicians. The
first movable-type printing press was invented not by Gutenberg but rather by Pi
Sheng in China around 1040. Paracelsus did not introduce the fourth element "salt"
and start the march towards modern chemistry, but a twelfth-century alchemist
from Kerala did so teaching in Saudi Arabia.

Many cultural communities made sophisticated astronomical observations
repeated only centuries later in Europe. For example, many of the observations that Galileo's telescope made possible were known to the Dogon
peoples of West Africa more than 1,500 years earlier. As Harding (1994)
notes, either they had invented some sort of telescope, or they had extraordinary eyesight.

At a minimum, these and related observations suggest that the development of science and technology was multicultural and that a great deal of
sharing of ideas took place. Hobson (2004) makes the stronger claim that up
until the beginning of the nineteenth century Europe was developmentally
backward compared to the East, and that Western civilization owes much
of its success to borrowing rather than innovation. For example, Francis
Bacon (1620/1960) suggested that the three most important world discoveries were printing, gunpowder, and the compass. All three were invented
in China (see Needham 1959 for a history of Chinese science).

Hobson provides an extensive historical analysis to back up his thesis.
Here are some examples with respect to science and technology. Chinese
per-capita iron production increased sixfold between 806 and 1078, to a
level achieved in Europe only by the nineteenth century. As late as 1588,
the largest English ships displaced a mere 400 tons; these were dwarfed
by the much earlier Chinese junks of over 3,000 tons. Financial practices
and institutions came from the Middle East and Persia. By 1730 Japan had
developed a futures market, something that was not seen in Europe until
more than a century later. The horseshoe probably came from the Huns
and stirrups from India and China. James Watson's 1776 steam engine
and Abraham Darby's coke-smelted cast iron (1709) were preceded by Wan
Chen (1313) in China. Vasco da Gama was not the first to sail around
Cape Horn—this was accomplished much earlier by the famous navigator
Ahmad ibn-Majid (and besides, da Gama used a Gujarati Muslim pilot). In
summary, Hobson argues that major innovations in Europe were imported
from elsewhere, especially the Middle East and China.

India was also an important influence. B. N. Seal's 1985 book *The Positive Sciences of the Ancient Hindu* is best known today for its chapter on the scientific method and its elaboration in the systems of Indian philosophy (*darsanas*). Raina (1997) reviews historical Indian analyses of criteria or tests of truth, perception, observation, experiments, fallacies of observation, the doctrine of inference, and determining causality.

Teresi (2002) has systematically documented what he describes as "lost discoveries," by which he means discoveries lost to historians of science. For example, modern evidence suggests that Copernicus developed his heliocentric theory by relying on a theorem borrowed from an Islamic astronomer, Nasir al-Din al-Tusi, who had lived three hundred years before. Although one could argue that this was a case of independent discovery, the evidence indicates that Copernicus's math contained many arbitrary details identical to al-Tusi's, undermining the notion of independent discovery.

Even ancient Greece took many of its ideas from Egypt. Teresi notes that the Greeks frequently acknowledged this debt but that revisionist histories associated with the nineteenth and twentieth centuries tend to filter out non-Western influences and omit this fact.

Contemporary Histories of Science

Hess (1995, 67) observes, "The very idea of the Scientific Revolution may someday come to be rejected as ethnocentric," as may the very idea of "Western science." In a similar vein, Burton (1995, 279) comments: "The colonisation of mathematics has been so successful that the history of their own mathematical culture and its contribution to knowledge is often unknown to students in Africa, Asia, and Latin America."

Although there may be particular points one could quibble about, the overall evidence is pretty condemning with respect to the idea that the West has an exclusive or even privileged position with respect to scientific progress. Galileo did not invent scientific research, Bacon did not establish the scientific method, and Newton was not the first to apply mathematics to observational data. Returning to Hobson (2004) for a moment, he offers the interesting observation: If the history of science were written around 900 when the Islamic Middle East was dominant, scholars might have speculated that Europe's cold temperate climate led to its people being backward, ignorant, and lacking scientific curiosity.

Science versus Superstition

Part of the argument for a unique, contemporary (Western) science has been that it requires a distinct, objective perspective, a perspective that is lacking in "less advanced cultures" where thought is dominated by magic, religion, and other forms of superstition. On this account, the advance of science critically entails overcoming irrationality.

Early research in anthropology tended to support the idea that superstition was pervasive. Nader (1996, 4) suggests that "anthropologists bounded science by comparison with magic and religion, probably because both magic and religion were potent and competitive with the science of their times." Frazer's (1890, 1912) multivolume *The Golden Bough: A Study in Magic and Religion* reviewed a large body of cross-cultural research and conceptualized a linear development from the age of magic to the age of religion and, ultimately, to the age of science. It documented a systematic influence of nonrational, magical beliefs on cultural practices. For example, Evans-Pritchard (1937) reported that the Zande believe that the application of fowl excrement is an effective treatment for ringworm, because they resemble each other. In other words, people described pejoratively by some as "primitive" may believe that causes and effects resemble each other and, in effect, conflate likeness and co-occurrence likelihood. Frazer also noted a related principle of magical beliefs: contagion, the idea that once two objects come into contact with each other they continue to affect each other.

One problem with this analysis is that these principles appear to hold equally for "modern thought." For example, Rick Shweder (Shweder et al. 1977) pointed out that (contemporary) research on personality also tends to confuse likeness and likelihood. In a similar vein Paul Rozin, Carol Nemeroff, and their collaborators have shown that the principle of contagion is alive and well in the United States, even among college students (Nemeroff and Rozin 1994; Rozin, Ashmore, and Markwith 1996; Rozin, Markwith, and Nemeroff 1992; Rozin, Millman, and Nemeroff 1986; Rozin and Nemeroff 2002). For example, people indicate that they would be very reluctant to wear a sweater once owned and worn by Hitler, no matter how extensively it has been cleaned. In short, if likeness and contagion are markers for magical thought, we are all guilty of it.

Assuming that none of us is rational 100 percent of the time, we can turn the question around and ask whether people in non-Western cultures do adopt a relatively unbiased, empirical stance often enough to acquire important knowledge about how the world works. One of the most important twentieth-century anthropologists, Bronislaw Malinowski, had no doubt on this point. He wrote:

There are no people however primitive without religion and magic. Nor are there, it must be added at once, any savage races lacking either in the scientific attitude or in science, though this lack has been frequently attributed to them. . . . On the one hand there are the traditional acts and observations, regarded by the natives as sacred, carried out with reverence and awe, hedged around with prohibitions and special rules of behavior. Such acts and observances are always associated with beliefs in supernatural forces, especially those of magic, or with ideas about beings, spirits, ghosts, dead ancestors, or gods. On the other hand . . . no art or craft however primitive could have been invented or maintained, no organized form of hunting, fishing, tilling, or search for food could be carried out without the careful observations of natural process and a firm belief in its regularity, without the power of reasoning and without confidence in the power of reason; that is, without the rudiments of science. (Malinowski 1925, 17).

In commenting on Malinowski, British anthropologist Edmund Leach (1957, 128–129) made the tongue-in-cheek comment, "Malinowski argued that primitives were just as capable as Europeans of making such distinctions. . . . He would have had a much better case if he had insisted that Europeans are ordinarily just as incapable as Trobrianders of distinguishing the two categories."

Brent and Elois Berlin and other ethnobotanists have argued that the medicinal use of plants in indigenous societies almost surely is based, at least in part, on explicit experimentation (B. Berlin 1992; B. Berlin et al. 1999; E. Berlin et al. 1996). It may not involve double-blind, randomized treatment studies with untreated control groups (a practice developed fairly recently), but it is experimentation nonetheless.

One Native American elder told us that his ancestors learned a lot about medicine from watching what bears did when they were wounded (applying parts of plants to their wounds) and then imitating them. In any event, the point we wish to make is that formal experiments with control groups represent just one tool out of a broad arsenal of methods for learning about the natural world.

More detailed cultural analyses go beyond broad claims of an empirical orientation to examine the frameworks in which research is conducted. For example, anthropologist Colin Scott (1996) has studied knowledge construction among the St. James Bay Cree and finds both similarities to and differences from the typical orientation of Western science. He writes (96):

If one means by science a social activity that draws deductive inferences from first premises, that these inferences are deliberately and systematically verified in relation to experience, and that models of the world are reflexively adjusted to conform to observed regularities in the course of events, then, yes, Cree hunters practice science—as surely all human societies do. At the same time, the paradigms and social contexts of Cree science differ markedly from those of Western science—accustomed as we are in the West to a "Root metaphor" of impersonal causal forces that opposes "nature" to "mind," "spirit," and "culture," and conditioned as we also are to view legitimate scientific procedure and production as the prerogative of particular professional and institutionalized elites.

He also notes that from a Cree framework for understanding (73),

human persons are not set over and against a material context of inert nature, but rather are one species of person in a network of reciprocating persons. These reciprocative interactions constitute the events of experience. . . . One consequence of this construction of the world is that an attitude of dogmatic certainty about what one knows is not only untruthful but disrespectful. There are many signs of recurrence and regularity in experience, but interpretations cannot be certain or absolute.

Scott's work is interesting because it examines Cree science on its own terms rather than using Western science as the sole standard. His analysis also presages several themes we will take up later on, such as the fact that scientific methods can be applied from different (epistemological) perspectives that may lead to different insights about the natural world.

Most cultural studies of scientific thought ignore one culture, the very one that brings us Western science. Though this has changed some since then, Nader (1996) comments that modern science is often treated by anthropologists as if it were beyond analysis. This neglect may reflect the implicit assumption that modern science has purged any cultural assumptions from itself. This assumption has not gone unchallenged (e.g., Viveiros de Castro 2010; Stewart 2010), and we will subsequently review different components of this challenge by a variety of scholars.

In summary, science and superstition do not divide cultures into distinct camps, nor do we suspect that they are neatly segregated either in individual cultures or individual minds. As Scott (1996) suggests, myth may embody differing orientations toward nature, and these differing orientations may affect what and how science is done. Mathematical analysis, skepticism, controlled experiments, hypothesis testing, and replications are science practices, but this list is far from exhaustive and ignores fundamental knowledge-producing processes such as observation, tool building, tinkering, and even daydreaming.

Science as Objective

In the final section of this chapter we take up the issue of the neutrality to science. We first consider the distinction between basic science and applied science and then turn to an analysis of basic science itself.

Protecting the Purity of Pure Science

In a brilliant public relations move, the field of science has tapped into the distinction between basic science and applied science. Basic science is scientific research conducted with the primary purpose of understanding how the universe works. Applied science takes advantage of basic science to focus on the use of knowledge for some practical objective such as curing a disease or technological innovations like cell phones or flat screen televisions. Of course it is misleading to imply a simple progression from basic research to application. Indeed, one of the strongest arguments for basic research is that it often leads to applications that could not have been foreseen in advance. A breakthrough in understanding and treating cancer may derive from a wide range of subdisciplines of science rather than only from "cancer research." This pattern of findings is so common that it is beyond dispute, no matter how often politicians may attack basic research based on the title of grant proposals (e.g., "Thermoregulatory Factors in Fruit Fly Reproduction").

The distinction between basic and applied research makes intuitive sense, so one might wonder why we call it a brilliant public relations move. Our answer is that this distinction also neatly allows science to shunt any issues involving values to the application side of things, so as to keep basic science pure. Thus a biologist might pursue the question of how cells work without any consideration of how the knowledge garnered by the research is used (e.g., to treat cancer or to build more potent biological warfare weapons). Scientists are also adroit in focusing on the uncontentious, positive applications of knowledge rather than more controversial uses of knowledge (e.g., building a bigger, better bomb, or developing delivery systems for biological warfare).

In short, strategically speaking, the field of science finds it desirable to celebrate value-neutral "knowledge for knowledge's sake" and to (selectively) point to applied science that improves people's lives. Sandra Harding refers to this as a triumphalist view of science. She says (Harding 2008, 4):

> By triumphalism I mean the assumption that the history of science (which, for triumphalists, is thus the exceptional history of Western science) consists of a narrative of achievements. For triumphalists, this history has no significant downsides. From this perspective, Hiroshima, environmental destruction, the alienation of labor, escalating global militarism, the increasing gaps between the "haves" and the "have nots," gender, race and class inequalities—these and other undesirable social phenomena are all entirely consequences of social and political projects.

Is it really the case that basic research and applied research are so easily differentiated? Although people commonly think of technology as applied science, the development of scientific tools like the telescope and spectrometer is celebrated as part of science itself. The U.S. goal of going to the moon was justified, in part, by the applications generated by needing to develop heat-resistant materials and miniaturized components. But was the goal in going to the moon to advance simple and straightforward basic science? One could argue that it was equally a political goal of demonstrating American technological and scientific superiority to the USSR and the rest of the world. It also might be seen as an applied goal, based on the assumption that basic science would be advanced by the goal itself. In our opinion the goal of going to the moon has basic and applied science (and policy) components so interwoven that the distinction between them is problematic.

One could argue that basic research and applied research exist on a continuum and that going to the moon is an unusual instance of the ambiguous middle ground. We make a ready distinction between day and night even though they grade into each other at sunrise and sunset. We believe, however, that the situation with basic and applied research is more akin to a normal bell-shaped distribution (see figure 2.1), where we can identify clear instances of theory and clear instances of application but most research occupies a multifaceted middle ground. For example, in the field of psychology three examples come quickly to mind: (1) the development of signal detection theory grew out of communication needs associated with World War II; (2) celebrated theoretical advances in the field of decision making such as prospect theory (Kahneman and Tversky 1979) have been grounded in everyday, economic decisions; and (3) basic research on text comprehension is inseparable from application to reading and learning. If basic research and applied research are joined at the hip, then it just is not credible to assign all value-laden components to the latter. If we haven't persuaded you on this point, the next section queries the value neutrality of pure, basic research.

Basic Science as Objective and Value-Neutral

Let us start with an anecdote from when the first author was serving on a committee whose job it was to write about science learning (we are deliberately being vague to maintain confidentiality). A subgroup of us wanted to include in our report the thesis that science is cultural in the sense that it reflects the culture and cultural values of its practitioners. Everyone on the committee seemed to agree, but for several iterations of our draft report this notion of "science as cultural" (somehow) mutated into the idea of a "culture of science" that would-be scientists need to learn. The various sciences do have consensual practices, so the idea of a culture of science is not incorrect; it just misses the point we were trying to make. It also implicitly endorses the idea that individuals somehow are able discard their cultural values at the doors of science. As Sandra Harding (2008, 4) puts it, "According to this view, only modern Western sciences have demonstrated that

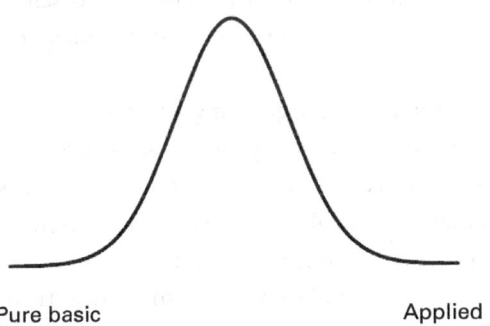

Pure basic   Applied

Figure 2.1
Hypothetical distribution of studies on a continuum of basic versus applied research. Pure basic and purely applied studies are rare, and most studies are a mixture of the two.

they have the resources to escape the universal tendency to project onto nature cultural assumptions, fears, and desires."

Indeed, routine practices associated with science are aimed at establishing neutrality. For example, the primate lab that the first author worked in during graduate school assigned numbers to monkeys rather than names, in an (unsuccessful) attempt to undermine any attachments being formed by the researchers.

Similarly, scientific writing often involves distancing devices such as the passive voice ("it is concluded that"). Longino (1987) comments: "Such language has been criticized for the abdication of responsibility it indicates. Even more, the scientific inquirer, and we with her, become passive observers, victims of the truth. The idea of a value-free science is integral to this view of scientific inquiry. And if we reject that idea we can also reject our roles as passive onlookers, helpless to affect the course of knowledge." And, we might add, we will see later on that distancing is not neutral, but rather reflects a perspective that influences scientific practices.

From one perspective, science is objective in that its goal is to describe how the world is rather than how we might like it to be. Bias in any form is eliminated because science must "grasp reality in its own terms." A politician may willfully distort information and be misleading and that may be good politics, but it is bad science. Some analysts draw a distinction between "Science" and "science." Capital-S science sets out the goals and ideals while small-s science refers to actual practices. Individual scientists may not be able to overcome their biases and some may even fabricate data, but science is self-correcting because it is a social activity and scientific skepticism and even professional rivalries tend to root out bias and deception.

One could accept this view and still use it to argue for greater diversity of scientists. The argument would be that science only can be self-correcting when individual biases differ enough to cancel each other out. If science is only done by white middle-class males, then whatever biases white middle-class males tend to share will be propagated, not eliminated.

Although we do not disagree with this logic, we want to make the stronger claim that diverse science is better science, even if there were no bias whatsoever. You might think of no bias as equivalent to "taking no position." We think a different metaphor for science is more apt—adopting a perspective. Imagine taking a photograph or drawing a picture (or even writing a story). Doing this reflects a perspective, and it is literally impossible to "take no position." This claim is central to our thesis and in the next chapter we will develop it further. Indeed, we will argue that in the absence of diversity, practices tend to develop that lead more narrow views to become more firmly entrenched and self-perpetuating.

It doesn't take a great deal of thought to realize that scientific practices embody, reveal, and reflect values. For example, in the United States animals are commonly used in research. To be sure, there are federal regulations forbidding unnecessary pain and suffering and otherwise specifying ethical guidelines for research with animals. When an author submits a paper to a scientific journal that includes research with laboratory animals, he or she must affirm that their research has conformed to these ethical guidelines. But different countries have different guidelines. In the aftermath of the use of the atomic bomb World War II, it was thought (notice how we use a distancing narrative here) important to understand the effects of whole-body radiation (as citizens of Hiroshima and Nagasaki had received) on human health and behavior. Would soldiers be able to fight effectively after being exposed to radiation? A number of primate researchers began to conduct studies on the short- and long-term effects of whole-body radiation on monkeys. When the government of India discovered that these studies were being done, they banned the export of monkeys to the United States.

Values are also reflected in the choice of research topics that are pursued. Traweek's list of accepted values included the idea that social, political, economic, and cultural factors affect which scientific ideas are funded or applied. For example, if medical research focuses on diseases that are common among white middle-class males and neglects diseases that afflict other groups, then if you're a member of one of these other groups, you can't be pleased. (We're not claiming this is true and the U.S. National Institutes of Health is proactive in countering any bias of this ilk, but it clearly is a relevant consideration. If you conceptualize AIDS as God's punishment for sinful behavior, you may assign less priority to funding research on AIDS as opposed to other diseases. Given that research is often costly, scientists do "follow the money.")

In a related psychological study Levin and Chapman (1993) found that (undergraduates') decisions about treatments for people with AIDS depended on whether they had contacted it from a blood transfusion or through risky behaviors. If you are a scientist who is morally opposed to birth control and see premarital sex as immoral, and further, if you have to decide whether to do psychological research either on why teens do or do not use birth control versus studying factors determining teen choices to abstain or not abstain from premarital sex, chances are you are going to focus on abstention.

Values may be implicit in science practices in a preemptive sense, if we allow the facts to precede questions about values. For example, now that the cloning of sheep such as "Dolly" has become established, it is much harder to raise the ethical question of whether mammals should be cloned. Harding (2008, 53) expresses this concern as follows:

It is the successes of science and engineering projects that have enabled scientists to convince themselves and us that they can predict and control nature. Consequently they think that they should have a monopoly on decisions about what constitutes reasonable standards for prediction and control, as well as about which social considerations are or are not relevant to such processes. They and they alone have the expertise to make such decisions, they claim. Thus modern sciences have been permitted to appropriate as purely scientific and technical matters political decisions about how we shall live and die.

The distinction between fact and value is also undermined by the very vocabulary research (must) use in describing phenomena. Descriptive concepts employed almost necessarily have both descriptive and evaluative dimensions. As Howe (2009, 430) observes: "Consider the concept of achievement. It is used to make value-laden descriptions; achievement carries a positive valence, unlike a pure descriptive concept such as the number five. Because such two-edged concepts are routinely (and unavoidably) incorporated into the descriptive vocabulary of social research, so, too, are the values of researchers, policy makers, and program designers participating in, sponsoring, or using such research."

Nor is this issue solely a problem for social sciences. Lewontin (1996, 297) in analyzing research in developmental biology comments: "To describe the life history of an organism as 'development' is to prejudice the entire problematic of the investigation and to guarantee that certain explanations will dominate." Later on (298) he elaborates, "Even molecular biology, with its talk of 'self-reproducing genes' that 'determine' the organism, is ideological in its implications. DNA is certainly not 'self-reproducing,' any more that a text copied by a Xerox machine is self-reproducing; in fact, it is the machine that is interesting and needs to be understood."

The final comment we'll make (for the moment) on values is that they are reflected in how a particular research topic is explored. Take a topic that is frequently in the news, U.S. students' lagging performance in science and mathematics. Even this description suggests that attention should be focused on what the United States is doing wrong, and one might further decide that the locus of failure is parenting, teaching, or some problem with the students themselves.

This deficit orientation may prove useful, but it tends to ignore the other side of this issue. That is, one could decide to study what other countries are doing right. If science and math learning is like figuring out how to build a fire (both literally and metaphorically), wouldn't you naturally pay attention to success rather than failure? The more general point is that almost all research topics can be approached from a variety of perspectives, and these perspectives are carriers of values.

Summary

It seems to us that the history of science that we learned in school is a clear case of European ethnoscience. If so, the first moral is that we need to acknowledge the contributions of many cultures to science and scientific methods. But the issue is not simply historical accuracy—values and fundamental orientations toward nature associated with scientific research are also at stake. Western (ethno-) science represents one set of orientations among many and one set of values among many. These values need to be made explicit and the possibility of a broader set of values acknowledged.

Similarly, "Science" will benefit from multiple perspectives and orientations. The next chapters further lay the foundation for these claims. Chapter 3 focuses on the role of models and theories in science. Although the "unity of science" is often taken as a given, we argue for a pluralist stance. Chapter 4 takes the perspective that science reflects the culture of its practitioners and describes some of the dimensions of science that are potentially culturally inflected. Chapter 5 describes some case studies of the cultural nature of scientific practices.

3   Maps, Models, and the Unity of Science

Scientists don't just reason; they interpret observations and experiments, they support or critique conjectures or hypotheses. They derive consequences, they extend models to new domains. They have multiple reasons for the particular choices and decisions they make in the course of all these activities, reasons that include feasibility, potential for application, aesthetic values, interest from other colleagues, interest from potential consumers, intelligibility to colleagues, resonance with metaphysical or ideological commitments. These are the kinds of factors under the umbrella of "the social." (Longino 2002, 98)

Science is science, right? There's just one science and you're either doing it by following the scientific method and the standards of science or you're not, and this doesn't depend on things like gender, ethnicity, or culture. Ultimately, it's the truth we're after, and our commitment to understanding how the world works means that we can replace less good theories with better ones and continue to converge on this truth. This is the position corresponding to what we call "the unity of science," and it is this view that we are aiming to challenge in this chapter.

But we are *not* signing on to what might be seen as the opposite view, "radical relativism," which we take to mean the idea that the "truth of the matter" is a social construction, grounded solely in consensus-building processes, much like the idea that "good art" is a social construction, as some might suggest. As we noted in chapter 1, there's actually a lot at stake here, and some scientists subscribe to the unity of science partly out of conviction, but also partly in the service of separating science from pseudoscience. Many scientists believe that so-called "creation science" is a pseudoscience, precisely because it does not follow the practices and standards of science.

What do we mean by "the unity of science"? Although there are a variety of scientific practices and methods that have evolved as consensual within various subfields of science, the key idea or principle is that there is "one best accurate" account of nature, the one that describes nature as it actually is. On the surface this idea may strike you either as obviously true or wildly implausible. On the side of implausibility, you might wonder what the behavior of quarks and other exotic particles could have to do with, for example, questions about how people make financial decisions or why the Itza' Maya of Guatemala categorize bats as birds. On the other side we have strong intuitions about realism, and it might seem intuitively plausible that the reason we can agree is that there's something to agree on, namely the world. From this perspective, the link between quarks and financial choices is an in-principle one. There are different levels of description and these two happen to be very far apart. For levels that are closer together the connections should seem more obvious—for example, do you think that the way the brain works has nothing to do with how the mind works? Not likely.

A good part of recent science history is precisely the interdisciplinary study of phenomena at different but related levels of description, as can be seen in the recent development of the subfields of neuroeconomics and social neuroscience. In short, the idea is that effective science often benefits from making connections among different levels of analysis, and it wouldn't work very well if theories of how the brain works were completely irrelevant to theories of how the mind works. Some scientists extend this argument to the idea of (again, in principle) "reductionism," the goal of explaining phenomena at one level of analysis in terms of processes taking place at a lower, more fundamental level of analysis. Reductionism presupposes the unity of science.

But embracing realism does not necessarily entail endorsing either reductionism or the unity of science. Here's an analogy we will develop in this chapter: Imagine some geographical area (e.g., Central Park in New York City) and then consider maps that we might make of it. A map is a representation that highlights some features and relations but ignores others. For someone whose only goal is to drive through the park it would be important to represent the roads that pass through it. One might even want a color code that represents the park roads that are closed on Sundays. A bicyclist would want a different map, though there would be relationships between these maps, as the roads closed to cars on Sundays are then (and only then) open to bicyclists. For a bird watcher the most useful map would reveal where different species of birds are likely to be found and perhaps include separate maps for different seasons. A police officer might prefer to have a representation of the location of any crimes that have been committed, perhaps with different codes for different times of the day. An entomologist would likely find none of these maps to be especially helpful, as they would likely represent the wrong scale, and a geologist wouldn't be pleased with any of these maps, because they have little or no topographic information, not to mention representation of subsurface features.

Even basic properties of maps like symmetry of distances (the distance from Point A to Point B is the same as that from Point B to Point A) might be set aside for certain purposes. For example, if the map represented not physical distance but rather walking times and Point A and B were at different elevations, then we might want one map for people at Point A going to B and a different map for people at Point B going to A.

The obvious point is that there are virtually an unlimited number of maps that could be drawn of Central Park. They would vary depending on the associated goals and purposes of the map users. They would all be grounded in the same reality, but often different aspects of this reality would be revealed or concealed by the alternative representations. The map of Central Park that a tourist might receive would reflect some guesses about what would be useful, as well as pragmatic constraints like portability—what tourist would want a map that was a mile tall, a mile wide (Central Park isn't square but you get the point), and weighed a ton?

Finally, any map we construct necessarily reflects a perspective on both scale and point of view. (And things like scale matter. A few years ago one of us visited the National Academy of Sciences Museum in Washington, D.C., where there was a map illustrating potential consequences of global climate change in Chesapeake Bay. The map necessarily was small and it made it seem as if a rise in the sea level of a meter or so would only have trivial consequences, contrary both to reality and to what the designers of the exhibit presumably intended to convey.)

Our discussion also has proceeded as if maps provide a (vertical) bird's-eye view that stops at ground level, which is what maps typically do. Note, however, that one could also have maps at (a horizontal) eye level, roughly as some GPS systems might, or maps that represent not ground level but rather geological phenomena at various underground depths.

In much the same way one can think of scientific models and theories as representing different aspects of reality and typically reflecting the goals of the inquiry. These theories are constrained by the ways things are in the world, but there may be an unlimited number of accurate descriptions, representations, and points of view, again depending on the perspectives and goals of the scientists.

Even the notion of accuracy can be something of a fudge factor in that it cannot necessarily be equated with "truth." As Longino (2002, 111) reminds us, "Boyle's law holds for ideal gases and often there are a number of qualifying clauses, etc. which means that the laws don't explain real world nonideal gases—but these representations serve as good guides and it seems we would want to call them knowledge even if they are not, strictly speaking, true." And later on (112), "Not only are laws, strictly speaking, false, but they are more useful in their false versions than in their indefinitely qualified, but true, versions." In other words, a model or theory often should be thought of not as true or false but rather as useful or not. Many fields of science employ mathematical models, and they necessarily always represent oversimplifications. As Godfrey-Smith (2006, 726) notes: "The modeler's strategy is to gain understanding of a complex real-world system *via* an understanding of simpler, hypothetical system that resembles it in relevant respects."

In this chapter we are going to focus on pluralism in three distinct senses: (1) pluralism in methods and approaches to research, (2) pluralism in levels of analysis in scientific theories, and (3) theoretical pluralism at the same level of analysis. In doing so our aim is to convey something of the social nature of scientific practices, and how that social nature helps undermine bias and increase the rigor of science. The various senses of pluralism help make the case for a diversity of perspectives and practices, a topic we will pursue in detail in chapter 4.

Pluralism in Practices and Perspectives

The Middle School on the Menominee reservation in Wisconsin has recently begun to have annual science fairs. The first author attended the 2010 fair and was impressed by the range and quality of the projects. Perhaps the most striking fact for Medin was that the projects were almost evenly split between two types: controlled experiments and design explorations (for the record, there were also projects based on observations that would fall into other categories). For example, one project compared the battery life of two brands of batteries to see which was better (controlled experiment). Another constructed three different bridges out of popsicle sticks to see which would prove to be most stable and support the most weight (design exploration).

Given Medin's training as an experimental psychologist in a field that almost exclusively focuses on controlled experiments, it was refreshing for him to see a broader picture of what counts as science. And perhaps one reason he was struck by these different types is that his academic appointment is split between psychology and education, and that also has implications for what counts as a good experiment. Many of his colleagues in education employ "design experiments," an iterative strategy of developing some educational intervention, and then analyzing, critiquing, and modifying it for another round of intervention and analysis (Bell et al. 2009; Edelson 2002). They conceive of these interventions as multifaceted and typically target some of these facets (but not others) for modification.

In psychology we typically think that (each of) our independent variables are (is) pure and homogeneous. For example, massed versus spaced practice, or subliminal versus supraliminal stimulus presentation, or stereotype threat present versus absent all seem like relatively straightforward variables compared with "inquiry-based" versus "control" science education. Of course, psychologists are deceiving themselves if they think of complex variables like culture as homogeneous independent variables.[1]

Medin's psychology colleagues would prefer a study where the intervention is contrasted with some control treatment and at the end of the study measures are analyzed to see if the intervention produces effects that are statistically reliable. These same colleagues might not think that design experiments are "good science."

By way of rebuttal, design researchers might, for example, point to the development of the telescope, which presumably happened incrementally, without the use of an experimental condition (the new telescope) and a control condition (the old one), though one could certainly do this sort of study. But when there is a clear theory of design (optics), there is no obvious need for a control condition (unless the quality of competing telescopes is in contention).

Education colleagues might offer a twofold elaboration (at least) of their rationale for design research: (1) in real-world school contexts it is difficult to find a well-suited control comparison and random assignment of students to conditions is not feasible (assigning one classroom to an intervention and another to a control raises the problem that any two classrooms might differ in numerous other ways unless the students were randomly assigned to them; even then, unless the same instructor teaches both classes, intervention versus control will be confounded with instructor; but if we control for instructor, time of day will be a confound); and (2) unless the intervention is optimal (a fact which is not in evidence), an overall contrast with some control doesn't allow you to identify the basis for success and failure in the way that a focus on the various components of a single intervention might. Ultimately, these differences in research perspectives are unlikely to be resolved through arguments and persuasion—instead they are more likely to be judged on pragmatic standards of success or failure in improving learning.

How does agreement get established within a subfield of science? Longino (2002) argues that it is a social process, and Traweek (1996) agrees that variables such as problem selection, "proper equipment," and adjudicating which experimental data to take as facts and which theories to take as important constitute a collective process. Traweek also notes that it is not an all-inclusive process and that scientists who are in power may determine who may be excluded from participating. Indeed, she suggests (133),

The definition of science is made by those who are empowered to offer resources for work they consider scientific; for example, the work funded by the NSF, SSRC, NIH, or NIMH is science [National Science Foundation, Social Science Research Council, National Institutes of Health, National Institutes of Mental Health]. These granting agencies set priorities for research based on assessments of national needs and the advice of scientists. These agencies draw on peer review to assess both the scientific merit and potential broader impacts of research grant proposals.

This may be a useful place to bring in the notion of "niche construction," a form of positive feedback in cultural and biological evolution. The idea is that organisms don't just adjust to their environments; they also often modify their environments in a way to increase their adaptive success. Plants may modify the soil in a way that increases their (and their offspring's) chances of survival, and politicians who get elected may change the voting rules to enhance their prospects of being reelected.

Niche construction is a straightforward, intuitive notion but it has very important implications for science and science-related practices. Naomi Quinn (2000, 139), commenting on niche construction by academic males, says, "There are obvious ways in which men designed academia to be an idealized men's world, from its reward of uninterrupted career trajectory to its reward of qualities (as risk, which I knowingly assume, of being called an essentialist) that academic women I know are, like myself, often not as good at—like bold assertiveness, selfish competitiveness, and individual glory-seeking."

Pierotti (2011, 198) further elaborates on niche construction favoring some over others in science when he writes: "The most likely explanation for the overall low numbers of scientifically trained Indigenous PhDs lies in the realization that the concepts and set of observations described as Western science is often considered to be a hostile environment for most Indigenous students." In short, the ideal science environment for one group of scholars may be counterintuitive and counterproductive for other scholars. Niche construction for one group can be niche destruction for another.

Let's return to our main story. In her book *The Manufacture of Knowledge*, Knorr-Cetina (1981) describes her studies of the sociology of scientific knowledge. For example, she traces a paper through its fifteen drafts in response to both internal reviews and external reviews associated with submitting the paper for publication in a scientific journal. The processes by which literature is created are also processes that shape what counts as good science. She also notes that the finished product is highly deceptive if taken as a historical record of the decisions made and actual work done from start to finish. (Note to our psychology colleagues: when you're writing a paper, you don't prepare two versions and ask other scholars to rate which of them is better. Instead, you follow the practices associated with design research.)

There's a risk that this is beginning to make science sound like judging jams and jellies at the county fair. In a sense this analogy is apt, in that judging at the county fair does involve standards for what makes a good jam or jelly. But it is also too much of a caricature because, at the end of the fifteen drafts Knorr-Cetina refers to, there will likely be a consensus that the final paper is far superior to the initial draft. In addition, scientific contributions typically become relevant when they are placed in the context of previous research—they rarely, if ever, represent isolated insights. To describe science as a social activity is to acknowledge that it is public and conforms to certain standards—in principle, your recipe is not a contribution to the science of jams if others cannot follow the same recipe and produce that same jam.

Even something as apparently simple as observation reflects pluralism, perspective, and social consensus. Quoting again from Longino (2002, 99): "On this account observation is not just seeing what is there but rather a social activity whose meaning may be negotiated. This includes decisions about what to observe, how to observe it and the significance of any particular observation." Haraway (1988, 583) sounds a similar note:

The "eyes" made available in modern technological sciences shatter any idea of passive vision; these prosthetic devices show us that all eyes, including our own organic ones, are active perceptual systems, building on translations and specific ways of seeing, that is, ways of life. There is no unmediated photograph or passive camera obscura in scientific accounts of bodies and machines; there are only highly specific visual possibilities, each with a wonderfully detailed, active, partial way of organizing worlds.

Let's pause for a minute to see where we are. First, there is little doubt that there is considerable pluralism in the practices across the sciences. To be sure, there are standards that hold across the sciences—science is public and observations by one scientist or a group of scientists must be repeatable by others. And while a poem or story may revel in contradictions, we want our scientific theories and observations to provide a consistent and coherent account. For example, one reason, if not the main reason, why scientists are skeptical about so-called paranormal or psychic phenomena is simply that they are inconsistent with a large body of other knowledge we have about how the world works. But even within a field like educational research, there may be sharp differences of opinion about research methods, and there are distinct communities of practice.

Perhaps we are also guilty of making science sound more chaotic than it actually is. One could argue that this messiness is a temporary state of affairs and eventually we'll get things sorted out. And when we do, we'll see the true unity of science. We are ready to take up this issue, and to do so we'll also need to discuss different levels of analysis in science and the important principle of reductionism.

Unity of Science, Levels of Analysis, and Reductionism

Some of our personal science heroes have expressed confidence in the unity of science (e.g., Wilson 1998; E. O. Wilson is a hero of Medin's for his fascinating studies of ants and wonderful writing, as in his 1992 book *The Diversity of Life*). Exactly what do we mean by "unity of science"? As we have been suggesting, the general idea is that there exists the potential for one consistent, coherent, true theoretical treatment of natural phenomena. That is, apparently diverse phenomena have discoverable relationships with each other and these relations can be described in a consistent, integrated way. According to Harding (2006, 137), "The unity thesis overtly makes three claims: there exists just one unified world, and one and only one possible true account of that world ('one truth'). And one unique science that can piece together the one account that that will accurately reflect the one truth about that one world."

For this thesis to get off the ground we're going to need two further ideas: (1) that there exist distinct levels of analysis in science and (2) that these levels are linked such that higher levels may be, in principle, explained by and are reducible to lower levels. For example, virtually everyone agrees that activities of the mind and of the brain are intimately related to each other. We might say that the brain is an "implementation of the mind" and note that there is a correspondence between electrical activity of the brain and our thoughts. When coupled with the further observation that distinct cognitive processes are associated with activity in distinct areas of the brain (and that we can measure this activity using noninvasive methods such as fMRI),[2] we have the basis for the exciting field of "cognitive neuroscience." The past few decades have seen an explosion of cognitive (and social) neuroscience research, based on powerful techniques for imaging brain activity. These developments constitute a scientific breakthrough, plain and simple.

Studying the relationships between mind and brain does not necessarily entail explaining higher-level processes in terms of lower-level mechanisms (reductionism), though it often invites it. In a popular article in the *New Yorker*, David Brooks (2011) claims, "Brain science helps fill the hole left by the atrophy of theology and philosophy." Similarly, Director of the National Institutes of Mental Health Thomas Insel (2010) proposes that neuroscience and its associated brain imaging offer the potential to see the "faulty circuits" that give rise to psychological disorders such as post-traumatic stress disorder or depression. Insel concedes that environmental factors may play a role in some mental disorders but nonetheless suggests that neuroscience is the most promising route to the prevention and cure of psychopathology.

Ideas involving the brain and reductionism can be seductive. Studies with undergraduates by McCabe and Castel (2008) and by Weisberg et al. (2008) have found that college students are more convinced by explanations that are accompanied by brain images and neuroscience language than by the same explanations not cloaked in neuroscience. It is reassuring that Weisberg et al. also found that neuroscience experts were not influenced by these uninformative accompaniments.

The language used in scientific articles, terms such as "the neural basis of" and "underlying," may invite explaining higher levels in terms of lower levels even when that is not the intention. For example, Murayama et al. (2010) describe research on brain activity and intrinsic motivation by saying, "These findings suggest that the corticobasal ganglia valuation area *underlies* [italics added] the undermining effect through the integration of extrinsic reward value and intrinsic task value." Without going into unnecessary detail, there is a body of research showing that extrinsic rewards can *sometimes* reduce intrinsic motivation to engage in some activity. Murayama et al. (2010) show that there are brain correlates of this undermining, but their data do not speak to the question of whether these are actually causes or effects—"concomitants" seems to be a safer descriptive language.

It is also easy to mix levels in the explanatory language employed. A recent report examining brain correlates of decision making involving the assessment of costs and benefits (Basten et al. 2010) was entitled "How the Brain Integrates Costs and Benefits during Decision Making." Is this different from asking how a person integrates costs and benefits during decision making? To answer this question, we would need a careful description of different levels of analysis.

The reason to take issue with the language employed is that it invites an asymmetry of causal explanation that may not be justified. Miller (2010, 723) comments, "rather than attributing mood changes to activity in specific brain regions, why not attribute changes in brain activity to changes in mood?" He cites (731) evidence to the effect that "psychotherapy appears to cause changes in EEG (Borkovec, Ray, and Stober 1998), that cognitive behavioral therapy normalizes hypoactive anterior cingulated cortex (Goldapple et al. 2004) and that medications and psychotherapy appear to have similar effects on PET-assessed brain activity (Baxter et al. 1992). Drug abuse clearly involves chemistry, yet psychotherapy has been found to be more effective than medication in treatment of drug abuse (Carroll and Onken 2005)." To this list we could add evidence that culture affects the brain (e.g., Park, Nisbett, and Hedden 1999).

The point of Miller's examples is that close correspondences between two levels of analysis do not imply that either is reducible to the other. He suggests (Miller 2010, 718): "A parallel could be drawn regarding social networking carried out by a network of computers. One's social network is not one's computer nor its connections to other computers. 'A map is *not* the territory it represents (Korzybski 1958, 58).'"

Pluralism and Levels without Reductionism: Marr's Levels of Analysis

David Marr was a vision scientist, but one of his contributions to cognitive science was to suggest that complex systems can usefully be described at three distinct levels of analysis, which he referred to as the (1) computational theory, (2) representation and algorithm, and (3) hardware implementation levels. The computational level asks what the goal(s) of the computation is and why it is appropriate, the second level asks how the theory and goals can be realized and how the input and outputs achieve them, and the third level asks how the algorithm is physically implemented.

To illustrate these levels, Marr (1982) offered the example of a cash register at a checkout, and we will both borrow from his example and update it. The broadest level of analysis asks what the system aims to do and why. In this case Marr's answer was that what a cash register does is addition. The theory of addition will describe what is being computed in a way that is independent of how numbers are represented (e.g., Roman versus Arabic numerals and how the addition process will actually be carried out). The reason that the theory of addition is relevant is that its properties conform to the rules for purchases (if I buy nothing, I should pay nothing; the total cost should not depend on the order in which purchases are entered or on how the goods are grouped; if I buy an item and return it, my net expenditure should be zero). (Note that our modern computational theory of cash registers would have to bring in further considerations. For example, cash registers are now used in inventory control, so not only the price but also the identity of the item being purchased is relevant to the inventory control system. The receipt that is printed is also itemized, which helps the purchaser check the price for accuracy.)

At the representation and algorithm level we would describe how the cash register (if we can still call it that, now that it is also a computer) carries out its operations. For example, the representation might be in the form of a bar code, which is then converted into Arabic numerals to perform addition. In states with a sales tax the cash register would also need to be able to compute the sales tax for just those items to which the sales tax applies. In general, there are a variety of combinations of representations and algorithms that could perform the same addition computations (e.g., it could be done in a standard decimal system or it might be done with a binary system and then converted to decimal numbers for ease of understanding by the purchaser).

Finally, the implementation level concerns how the addition (algorithm) is physically realized. In some old cash registers (older readers may remember Monroe calculators), addition was accomplished mechanically using gears that rotated numbers. In 1982 when Marr's book appeared, cash registers were virtually all electrical and used wires and transistors. And as we mentioned earlier, today's cash registers are virtually computers, and what the purchaser sees as their output (the printed receipt) may be input to other computer programs that analyze purchasing patterns. So computational systems often have multiple goals and functions.

Of course, when it comes to complex computational systems we're far more interested in human beings than cash registers. In the case of our visual system, presumably the computational theory entails having a representation of the world so that we can move about, avoid dangers, and so on. Vision science then attempts to describe what sorts of analyses might be useful (for example, what information we use to compute depth) and how the visual system is actually implemented (e.g., the role of lens, cornea, retina, and various brain regions).

Marr's framework is useful in that it directs attention to different kinds of problems or issues a complex system may face (with corresponding implications for different questions a theory might wish to address). For example, a critical question in vision science is just what information is available in the world for our visual system to pick up. Another question is how a two-dimensional projection of light on the retina is converted into our perception of a three-dimensional world. Note also that each level may inform other levels. If you don't know what is being computed, it may be difficult to tell how some computation is implemented in brains.

We should also add that there is nothing unique about having exactly three levels. A broad computational-level goal may be broken down into various subgoals, each with its own representations, algorithms, and implementations. For example, the goals and computations associated with motion perception may differ from those of color perception, which in turn may differ from those of face perception. Cognitive scientists Steve Palmer and Ruth Kimchi (1986) suggested that any complex informational event at one level of description can be specified more fully at a lower level of description by decomposing it into a number of components and processes that specify the relations among these components. For example, text comprehension can be decomposed into sentence comprehension, word recognition and comprehension, and even letter perception.

Both Marr as well as Kimchi and Palmer stress the benefits of conducting research and analyses using multiple levels of description, but they do not endorse reductionism in the sense of assuming that the most specific level of description is the best or correct one. Instead, different levels may address different questions, as in Marr's analysis, and Palmer and Kimchi stress that there may be emergent properties at higher levels of analyses. For example, the state of being liquid, solid, or gas is not a property of any single molecule but is instead a property of aggregates of molecules (see Chang 2008 for some of the complexities involved). These same issues play out when economists worry about relationships between macroeconomics and microeconomics and when cognitive scientists debate whether culture is an aggregate of individual beliefs and practices or something that transcends individuals.

Unlike Marr's levels, the Palmer and Kimchi framework does not entail qualitatively different levels. Nonetheless considerations like efficiency might determine which level is most useful. For example, Medin remembers the early days of smallish (everything is relative: these computers were about a meter wide and two meters tall, substantially smaller than the old IBM 1620 computers but vastly larger than today's machines) but not yet personal computers when researchers had to "toggle in" each machine language instruction to construct a program to run an experiment. This was soon followed by assembly language programming and compilers which resulted in enormous time savings. Nowadays one only needs to set a few parameters in some high-level programming language to run studies that would have taken weeks to set up in machine language. For purposes of programming and understanding what a program is doing, a programming language more abstract than machine language is the obvious choice. In our view this is one of the most powerful arguments against reductionism as a research strategy—it often is brutally inefficient, even if it were in principle possible.

A final positive feature of Marr's analysis is that it may protect us against mixing questions and assumptions across levels. To illustrate this issue we offer two examples, decision making and evolution. Arguably, the computational goal of a decision-making system is to maximize gains and minimize losses so as to achieve the best overall value or utility. It does not follow, however, that the best way to make decisions is to estimate costs and benefits and select the option that maximizes anticipated utility (Bennis, Medin, and Bartels 2010). The problem lies with the terms "estimate" and "anticipated," because there may be unforeseeable costs or benefits, and anticipated utility may depart from experienced utility. Hence, it is possible that the best way to make certain types of decisions might be to follow the advice of an elder, consult some principle or rule, or simply do what you did in a similar previous situation that turned out well.

The same consideration holds for evolution. At the computational theory level the goal may be to compete successfully to insure survival and reproductive success (we're ignoring questions about whether this competition is at the level of genes, individuals, species, or other units or combinations of units). But it does not necessarily follow that the most effective way to do this (Marr's representation and algorithm level) is through means of direct competition. Instead the most effective strategy may depend on a range of environmental factors as well as the set of strategies that other players adopt (see Bednar and Page 2007). Cooperation in one context may be the most effective strategy for competing more globally.

Levels of Analysis: Summary

This may be more time than you ever wanted to spend thinking about levels of analysis and reductionism, especially given the simple conclusions we hope you've come to: (1) it can be enormously productive to consider multiple levels of analysis and their interrelationships (one salient example being cognitive neuroscience), and (2) attending to multiple levels of analysis entails no commitment whatsoever to reductionism.

Although reductionism is one way to assert the unity of science, does giving up on reductionism entail giving up on the unity of science? It depends. One could define unity of science in terms of there being one best, true account of how the world is, but add the idea that this one account is comprised of a series of best, true accounts corresponding to different levels of analysis. This would allow us to preserve the ideal of one true science, reflecting only how the world is and not who is doing the research. This assumes, however, that there is some optimal number of levels of analysis and that within a level there is one best account. We now question these assumptions.

Pluralism within Levels of Analysis

Let's start with an oversimplified example. Imagine that you are a psychiatrist and that you have two treatment options you can use, cognitive-behavioral therapy and drug therapy. Suppose we also know that some patients respond well to cognitive-behavioral therapy and some do not, and that the same holds for drug therapy. The crucial question is what the relation is between responding to one treatment and responding to the other one.

First let's consider two special cases. It could be that all and only the people who respond to cognitive-behavioral therapy also respond to drug therapy. Another (equally unlikely) possibility is that patients who respond to one therapy do not respond to the other but all respond to one or the other.

Now imagine that we are developing an assessment system that we want to use to predict whether or not a given therapy will prove successful. For the two cases described, we would be pleased to have a single taxonomic system, in the first case distinguishing between responders and nonresponders, and in the second distinguishing between cognitive-behavioral therapy responders and drug therapy responders.

If we move beyond these special cases, the classification task becomes more difficult. Suppose that there tends to be a positive correlation between responding to the two forms of therapy but that there are a fair number of patients who respond to one therapy but not the other. We illustrate this in figure 3.1, according to which we assume that we have patients' scores on two assessment scales (the two dimensions of the figure), that a letter pair corresponds to a patient, and that a capital letter indicates that the corresponding form of therapy (B = behavioral, D = drug) will be effective. A little reflection will reveal that there is not a single perfect classification system, but rather that the optimal classification cutoff for cognitive-behavioral therapy (shown in figure 3.2) is different from the optimal classification scheme for drug therapy (shown in figure 3.3). It should be obvious that the best diagnostic criterion for purposes of cognitive-behavioral therapy is not the same as the best for drug therapy.

In general, there is no guarantee that the classification system that works best for one form of therapy is the best one for another form of therapy. At least with our hypothetical example and assessments in figure 3.3, we will need not one but two classification schemes. That is, we will need pluralism.

Both forms of therapy in this example focus on change taking place within the individual. Community psychologists may be more inclined to analyze the environments in which people live. Medin remembers a community psychologist colleague relating a story from his days at Yale University about having many young working mothers as patients and coming up with the treatment of organizing the community to provide better day care.

Is this example of alternative categorization schemes for different therapies at all realistic? We think that it is for the case of psychodiagnostic classification. The diagnostic system currently most widely used in the United States, developed by the American Psychiatric Association, is DSM-5. Presumably, it reflects in part a compromise document that balances competing research and psychotherapeutic interests, but we'll ignore that for the

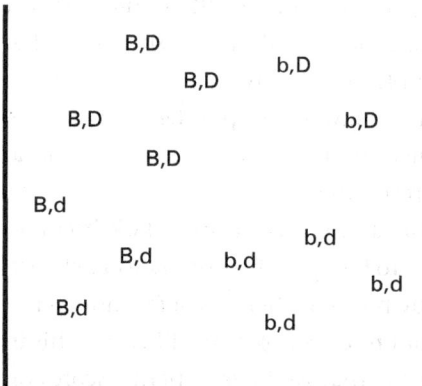

Figure 3.1
Relation between scores on two assessment dimensions and success (capitalized) or failure (small letter) of drug (D, d) or behavioral (B, b) therapy.

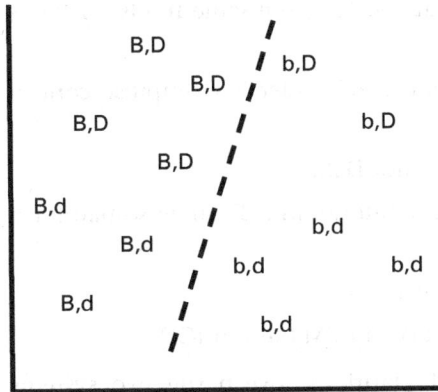

Figure 3.2
Decision line for assessment that allows one to predict who will benefit from cognitive-behavioral therapy.

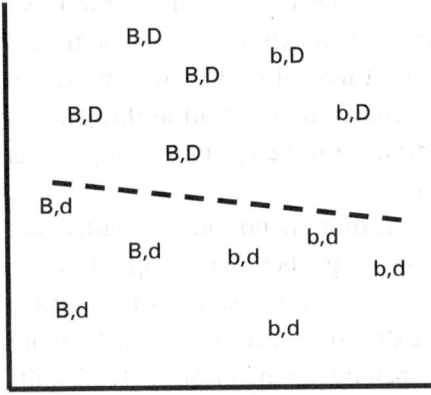

Figure 3.3
Decision line for assessment that allows one to predict who will benefit from drug therapy.

moment. For now, we just note that there is a similar, alternative system developed by the World Health Organization, ICD-9-CM. ICD is similar to DSM, but not identical. Here are some of the main differences:

1. Culture-specific disorders like (running) "amok" are in ICD but not in DSM.

2. A subtype of anxiety disorder, hysteria, is in ICD but not DSM.

3. Asperger's syndrome is in DSM but not ICD; the same holds for Rett's disorder.

4. Trichotilomania is in DSM but not listed under ICD impulse control disorders.

5. Psychogenic vomiting is in ICD but not DSM.

6. Body dysmorphic disorder is in DSM, but not in ICD under somatoform disorders.

7. Transsexualism is in ICD but not DSM.

8. "Acculturation problem" is a category in DSM but not ICD.

In short, there are very many similarities between the two systems but there are also some differences. Interestingly, psychotherapists in the United States use DSM for diagnosis but then, for purposes of submitting claims to insurance companies, are required to use ICD. This is pluralism both in theory and in practice.

Competing interests and perspectives may be in play, both here and elsewhere. Phenomena such as altered states of consciousness may be treated as symptoms of psychopathology by most mental health professions, but may be a rich source of ethnographic information for an anthropologist (e.g., Ward and Kennedy 1994). A medical anthropologist may adopt either perspective, depending on the context.

Laypeople tend to think of scientific taxonomy not only as settled, but also as objective and given by the natural "gaps" between groups of similar biological kinds. This is far from the truth, as the history of the conflict between competing approaches to classification reveals (see David Hull's 1988 book *Science as a Process* for a compelling narrative). Simplifying quite a bit, classical taxonomy focused on grouping organisms by shared properties such that organisms within a group were more similar than organisms belonging to different groups. This orientation has been replaced by cladistic taxonomies that focus on patterns of descent. By these criteria, dinosaurs are closer to birds than they are to mammals, even though dinosaurs may appear to be more similar to mammals than to birds.

Of course, whichever approach one takes, the answers may depend on what you choose to measure. Lloyd (2010, 205) remarks:

Different criteria can be used to decide which groups of populations should be accorded species rank (Mayr 1957). Among those favoured in the past have been (1) morphology, (2) cytology, (3) interfertility, and (4) the extent of DNA hybridization, while nowadays, taxonomists concentrate rather on (5) the analysis of SSU rRNA (small subunit ribosomal RNA) gene sequences (e.g., Simpson and Roger 2004). None of these is a merely arbitrary procedure: in each case, there can be more and less accurate applications of the criterion in question. But while all can be said to offer the possibility of yielding objective results, these do not all converge. To arrive at an orderly taxonomy the similarities and differences have to be *weighted* and this risks circularity: you get out what you put in. There is no one definitive zoological taxonomy since there is no neutral way of determining which criterion or set of them to invoke.

(Please forgive one more sidebar. The development of "scientific" taxonomic systems was dramatically accelerated during the colonialist era when biologists traveled the world and brought back plant and animal "specimens" for classification purposes. Given that these samples were taken out of their normal settings, the resulting taxonomic system necessarily gives little or no weight to relationships, ecosystems, or contexts. A classification system that considers these factors might look quite different and yield different insights.)

We return to psychology for a second example of pluralism at the same level of analysis. In this case the domain is causal reasoning. Broadly speaking, there have been two major theoretical approaches to understanding cause. One focuses on patterns of covariation and counterfactual reasoning (Gopnik and Schulz 2004; Gopnik et al. 2004; Griffiths and Tenenbaum 2005). If people who smoke cigarettes are more likely to get lung cancer than a comparable sample who do not, then smoking is a plausible candidate as a cause of lung cancer. Likewise, if someone drops a wine glass and it breaks, one might decide that the dropping was the cause of it breaking and reason that if the glass hadn't been dropped it would not have broken. One could even set up an experiment where we randomly assign wine glasses to be dropped or not dropped and record whether or not they break, and thereby confirm the accuracy of our counterfactual reasoning by the presumed covariation between dropping and breaking.

The second major framework for causal understanding relies on notions about mechanisms and a physical transfer of some entity from the cause to the effect (e.g., White 1999; Wolff 2007). For example, in the case of cigarette smoking, even people who do not know the exact causal mechanism are likely to reason that something in the cigarette smoke that comes into contact with the lungs is responsible for lung cancer. In the case of physical motion where, for example, a billiard ball is set into motion and squarely hits another ball, whereupon the first ball stops and the second moves away, people may have the impression that some force or impetus has been transferred from the first ball to the second one (even though we may learn in physics that the force applied by the second ball to the first is equal to the force applied by the first to the second).

Lombrozo (2010) recently has suggested taking a pluralistic approach to these two broad frameworks rather than seeing them simply as competing theories. Keil (1992) has noted that there may be distinct explanatory perspectives or "modes of construal," including the physical, teleological, and biological. For example, one may view human beings as physical objects when reasoning about whether a beach ball lobbed in their direction is likely to pass through them; as biological systems in wondering whether the germs on the beach ball might be a hazard; and as psychological or intentional agents in wondering why the person lobbed the beach ball in the first place.

Lombrozo proposes that what is true for explanatory frameworks is also true for causal ascription. She suggests that both counterfactual and mechanism frameworks are important but that counterfactual reasoning is more predominant where issues of intention and purpose are in play. Lombrozo (2010) also provides some empirical evidence that people assign more importance to counterfactual explanations when some outcome comes about accidentally rather than intentionally.

Issues of pluralism within the same or adjacent levels of analysis are especially important in interdisciplinary research. In one informative analysis, Longino (2006, 102) describes four approaches to the study of the development of behavior: (1) behavior genetics, (2) developmental systems theory, (3) neurophysiology, and (4) anatomy (neurobiology). Her review suggests that writers are very aware of the different approaches (but believe that only theirs is correct). These approaches also vary in the issues they choose to study and those they choose to ignore (e.g., gene-environment interactions). Ultimately researchers from these different orientations may be asking different questions.

Pluralism of Perspectives

We end this chapter by noting that pluralism in science should not be a source of distress or like two umpires disagreeing about a call. "Cultures are 'toolboxes,' not just 'prison houses,' for the production of knowledge" according to Harding (2006, 153).

Ronald N. Giere (2006) offers an analogy between multiple perspectives and the variability of color vision both within and across biological kinds. There is not one form of color vision (or its absence) that is "correct," just differences that have evolved under different conditions. For human beings, color vision is rich and allows the detection of differences that color-blind people do not see. But even color blindness offers some advantages in detecting camouflaged targets, as we learned in World War II.

More generally, different perspectives typically offer different kinds of pluses and minuses. Some subareas of biology rely heavily on "model species" that have been carefully selected and developed to provide experimental controls or for ease of observation and analysis (e.g., fruit flies for studies of genetics because their life cycle is short, or the squid axon because it is so large). The cost is that convenience and control may be being traded for representativeness and generalizability. In some cases, research may use lineages found only in laboratories and it may be challenging to extend findings from the lab studies to the field.

The final example of perspective we examine is an analysis of the role that metaphors and broad explanatory orientations may play, in this case in the domain of biology (Levins and Lewontin 1985). Levins and Lewontin suggest that the field of evolutionary biology reflects theoretical assumptions that are culturally inflected. For example, the focus on individualism and the idea of progress (even Darwin was not immune to the notion of steady improvement in an absolute sense: at the end of the *Origin of Species* [Darwin 1872/1998, 489], he writes, "And as natural selection works solely by and for the good of each being, all corporeal and mental endowments will tend to progress toward perfection") may lead researchers to make a sharp division between organisms and their environment, with the latter being treated as passive and fixed. They note that this assumption is problematic if it leads researchers to ignore the fact that organisms may both select the environments they occupy and modify their environments (as we mentioned earlier, the term for this is "niche construction"). As they say (Levins and Lewontin 1985, 99), "The environment is not a structure imposed on living beings but is in fact a creation of those beings." For example, grazing animals actually increase the rate of production of forage, by fertilizing the ground with their droppings and by stimulating plant growth by cropping.

We take these observations to reinforce the value of multiple perspectives in scientific endeavors. Cultural diversity is one means of reinforcing pluralism in perspectives. We believe that researcher diversity broadens the identification and appreciation of epistemically significant questions.[3]

Summary

Although many scholars argue that the issue of pluralism versus the unity of science is ultimately an empirical question, we find the case for pluralism to be compelling. Pluralism thrives on diverse perspectives and cannot survive the "one true way" orientation that is presented in some popular representations of the nature of science.

Another conceptual limitation of thinking of science as truth is that it ignores the fundamental role of models and theories in science, and models and theories necessarily make simplifying assumptions or correspond to idealized conditions that are never actualized in nature. Furthermore, theories and models are constrained by the fact that, in our case anyway, human beings must be able to understand them and this also requires simplification and idealization. But again there isn't necessarily just one possible simplification, and different models in the same domain may be useful for different purposes.

We think that Sandra Harding (2006, 141) nicely summarizes the argument when she concludes, "the goal of only one true account of the natural world seems too little to ask."

4 Values Everywhere within Science

The roads to human power and to human knowledge lie close together and are nearly the same. (Bacon 1620/1960, 122)

This sentence from Francis Bacon can be read a number of different ways, the most innocent being that knowledge is power. A more controversial understanding is the claim that those in power have control over the knowledge that gets produced. This implies that knowledge, far from being innocent, can be a tool for domination, used to support those in power. In the words of Charles Schwartz (1996, 148): "The work of science, when it is not just an exercise in idle curiosity by some individual, is generally integrated into the social and economic structure of society. That is to say, science is not separated from politics. In a world where knowledge is power, one should thus expect that the activities of science are largely under the control and direction of those sectors of society that hold dominant political power."

The one thing that is completely unambiguous is that Bacon would have rejected the argument that knowledge and power have nothing to do with each other. But that is often precisely the claim of those in power, including prominent scientists.

Aside from explicit racism and domination, why should anyone think that knowledge and power are unrelated? While there may be many reasons, we want to highlight two factors motivating this view. One is blindness to privilege and a tendency to see other people who disagree with us as biased while we remain objective. It is very easy to equate "how we do things" with "how things are done," and to see deviations from this as deviations from an objective norm.

The second factor is the notion of "knowledge for knowledge's sake," a deep yearning to understand the natural world and a curiosity that can be all-consuming and transparently "innocent" of any considerations of power. Brian Yazzie Burkhart (2004) poignantly describes how this move reflects explicit and implicit values that may be markedly different from those of indigenous peoples who "must maintain our connectedness, we must maintain our relations (to the lifeworld), and never abandon them in search for understanding, but rather find understanding through them."

This chapter will review some of the ways in which values are embedded in science practices. We start with an extended quote, one that is a bit contentious but nicely summarizes the perspective we are advocating:

We cannot restrict ourselves simply to the elimination of bias, but must expand our scope to include the detection of limiting and interpretive frameworks and the finding or construction of more appropriate frameworks. . . . Instead of remaining passive with respect to the data and what the data suggest, we can acknowledge our ability to affect the course of knowledge and fashion or favor research programs that are consistent with the values and commitments we express in the rest of our lives. From this perspective, the idea of a value-free science is not just empty, but pernicious. Accepting the relevance to our practice as scientists of our political commitments does not imply simple and crude impositions of those ideas onto the corner of the natural world under study. If we recognize, however, that knowledge is shaped by the assumptions, values and interests of a culture and that, within limits, one can choose one's culture, then it's clear that as scientists/theorists we have a choice. We can continue to do establishment science, comfortably wrapped in the myths of scientific rhetoric or we can alter our intellectual allegiances. While remaining committed to an abstract goal of understanding, we can choose to whom, socially and politically, we are accountable in our pursuit of that goal. In particular we can choose between being accountable to the traditional establishment or to our political comrades. (Longino 2002, 8)

Bias in Science

The goal of scientists to be unbiased is an ideal that is sometimes violated. There are cases of scientists "fudging" data, perhaps to convert some trend into a difference that is "statistically reliable." It's a risky thing to do because scientists fall under suspicion when other scientists fail to replicate their findings. But we know that it happens because some scientists have been caught at it and confessed.

Shortly after Medin took his first academic job at Rockefeller University, Frederick Seitz, a nuclear physicist, became the president of the university. Seitz went on to be president of the National Academy of Sciences and was one of our country's most respected scientists. Only recently did Medin find out that Seitz also was a powerful advocate of the view that smoking does not cause cancer. Seitz then went on to claim that secondhand smoke was not a health hazard, next that the ozone layer was not at risk, and finally to the position that global climate change was a fiction (see Oreskes and Conway, *Merchants of Doubt*, 2010, for this and related stories of questionable practices by scientists). It is difficult to escape the conclusion that Seitz let his political values and orientation (as well as the remuneration he received for taking these stances) have greater priority than his concern for truth. He sought to obfuscate rather than clarify.

There's no place in science for this sort of bias. If a scientist is paid to produce a finding, or lets the desired results supersede how the results actually are, then it's plain and simple bias, regardless of whether the underlying motivation is selfish or unselfish. As we shall see, however, values influence science even when bias is eliminated.

Values in Different Forms

It may be useful to make a few distinctions before continuing our discussion. The first is between contextual assumptions associated with science practices and constitutive assumptions that some argue form the core of science. Constitutive assumptions include notions about science being public (no one could get far by claiming that they have a secret way of turning rocks into gerbils), focused on findings that can be verified, and adopting procedures for supporting claims and arguments (e.g., replication, statistical tests, ruling out alternative interpretations). Contextual assumptions would include decisions about what to study, how to study it, what to measure, and the like.

The second and related distinction is between the context of "discovery" and the context of verification. The context of discovery includes things like coming up with research ideas or generating hypotheses; verification draws on the constitutive assumptions of science to establish the value of the discovery. The context of discovery is much less constrained than the context of verification. It is fine for a scientist to have a theory or idea come to her in a dream, especially when it is supported by other analyses and experiments. But it just won't work for a scientist to dream that her theory is supported by other observations and conclude from this that her theory is correct.

Discussions of science and the scientific method tend to focus on constitutive assumptions and the context of verification rather than contextual assumptions and the context of discovery. Graduate students may receive advice about how to generate good ideas or how to tackle some research problem, but this sort of discourse is largely absent from undergraduate courses. Similarly, statistics courses concentrate on what to do with the numbers that come from some observations or experiment and say little or nothing about how to decide what to measure or where the numbers come from in the first place. But it is in just these sorts of considerations that values may be most directly reflected.

Values Embedded in Scientific Research

We now turn to a description of some of the many ways in which values are infused in scientific practices. Our examples draw heavily on social, behavioral, and educational sciences because that is where our expertise lies. (See also Solovey 2013 for an analysis of interactions among money, politics, and social sciences.) For other examples, we'll mainly quote other scientists with relevant expertise. As we will see, variations in particular values are often correlated with characteristics of the scientists themselves.

Decisions about What to Study

Imagine that you are an educational researcher and that you want to improve children's learning in schools. You might decide to do research comparing successful learners with less successful learners, to identify ineffective learning strategies or practices that get in the way of success, and then develop some intervention to see if you can change student strategies to improve children's academic success. You could equally well focus on teacher success and failure in fostering student achievement. At some point you might even look for interactions between teacher qualities and student qualities and success, with the idea that achievement might be improved by matching students' tendencies for learning with teaching practices.

Although it seems natural to focus on students and teachers, there are other factors that may be worthy of attention. For example, environmental variables may be important. Having assigned seats that are arranged in rows may produce different patterns of learning than having clusters of tables. A sparsely furnished classroom may help some students concentrate, but others may profit from a richer, more stimulating environment.

And cultural assumptions may have a large influence on the kinds of studies we do. For example, it is normative in the United States to focus on individual achievement and to assume that the individual student is the relevant unit of analysis. Imagine that U.S. mainstream culture valued group accomplishment much more than individual success. In this case we might take collaborative teams as our unit of analysis and focus on team members' interactions and coordination strategies. (Here's a possible case in point. When Medin's granddaughter was in fourth grade he went to visit her classroom. As was the case when Medin was in fourth grade, there was a chart reflecting the amount of reading being done, including a "thermometer" reflecting progress toward the reading goal. In Medin's classroom each student had a column that increased with each book he or she read. But what was being charted in his granddaughter's classroom was the total class reading results, not individual accomplishments. Obviously, it would take further classroom observation to see if this practice was part of a pattern of encouraging collaborative goals [Medin is willing to volunteer for this task], but it certainly is in contrast to the individualistic focus when Medin was in grade school.)

Scheman (2003) claims, correctly in our view, that there has been a corresponding individualist assumption as a guiding force in the field of psychology. She says (226), "The largely unquestioned assumption, that the objects of psychology—emotions, beliefs, intentions, virtues and vices—attached to us singly (no matter how socially we acquire them), is, I want to argue, a piece of ideology."

What to study is also driven by intellectual interest and social reinforcement. Laura Nader (2010, 264) summarizes analysis of responses of physicists and engineers to her 1981 *Physics Today* article critiquing the values of physicists as follows:

A summary statement of my observations would include the predominance of group-think; people who thought differently than the group were told they were off the track. Part of group-think was a lack of respect for diverse solutions and for diverse kinds of intelligence, an avoidance of technologies other than nuclear or coal, and a preference for abstract rather than concrete thinking (as if figuring out how to dispose of waste was equivalent to actually doing it). Memos discussed nuclear, coal, and nonnuclear energy. Nonnuclear meant solar, which was described as "an orphan child," "not very intellectually challenging" and "just a bunch of mirrors." . . . In addition, I tried to understand the concept of the Big Toy. In answer to my question, "Why did you go ahead with nuclear energy without having first solved the nuclear waste problem?" I heard "because it is interesting, it is fascinating, it is fun" even though the same people would unselfconsciously agree that such a direction takes us into unknown and dangerous waters.

Decisions about the Framing of Research Questions

What appears to be the same research issue or scientific question often may be studied differently depending on how the question is framed. People in general and scientists in particular typically frame questions in terms that relate the new to what is already known. If some novel situation arises, we ask how it relates to what we already know rather than how what we already know relates to what is new. Similarly, we ask why things deviate from a norm rather than why they conform to some norm. For example, researchers are far more likely to study why students drop out of school than to ask why they stay in school. These variants may sound like the same question, but they are not.

Here's a related example. Nisbett and Cohen (1996) were intrigued by the question of why the murder rates in the South of the United States are much higher than in the North. They drew on an impressive range of statistics, control variables (e.g., population density), observations, and even experiments to identify the "culture of honor" in the South as the likely key factor. Note, however, that if their motivating question was why murder rates in the North are so much lower than those in the South, their analysis would likely have focused on the North and the conclusion might have taken a quite different form. We doubt, for example, that the conclusion would have been that the reason murder rates are lower in the North is that the North lacks a culture of honor. It is not that one conclusion would be wrong and the other right; instead, different framings may produce different insights.

Continuing with this example, we also note that the framing of a research question dictates its scope. One could have begun with the question of why the South not only has higher murder rates, but also greater rates of military service and sacrifice for one's country than are seen in the North. One might still have been led to an interpretation in terms of a culture of honor, but the broader scope of including military service undermines the implication that a culture of honor is uniformly bad (we hasten to add that Nisbett and Cohen did *not* draw this conclusion).

Influences of Guiding Metaphors

We've mentioned evolution and competition before, but we bring it up again to make a point. Gross and Averill (2003, 71) suggest: "Nature, as depicted in biological science, is a man's world. For researchers inevitably project the visions their imaginations, and the attitudes their life experiences make available and most biologists have been men." They argue that this has led to an undue focus on scarcity, struggle, and competition and note that varying the focus of research might well yield different insights. Gross and Averill ask (85): "Why not see nature as bounteous, rather than parsimonious, and admit that opportunity and cooperation are more likely to abet novelty, innovation, and creation than are struggle and competition?"

   If you don't mind, you can participate in an informal study. One important aspect of human evolution is the development and use of tools. This doesn't come as news to you, but we ask you to visualize a prototypical example of making and using a tool.

   If you're like Medin, you thought of something like an ax, knife, or some kind of weapon and you imagined the creators to be men. But suppose you think of nature as a woman's world. Now you might consider the possibility that women were responsible for breakthroughs in the development of tools (e.g., for agriculture, cooking, sewing). In a tongue-in-cheek analysis Hubbard (2003) dismisses this form of male bias by asking the ironic question of whether only men have evolved.

Orienting Assumptions Provide Relevant Units of Analysis

You are now invited to do another experiment. When you have a moment go on line and enter "ecosystem" on Google Images. We're betting that the most common depictions do not include human beings. Those that include humans are likely to have them at the top of something like a food chain. This positioning is itself a limiting metaphor as it ignores the literally trillions of "lower organisms" that make our life possible and break down our bodies when we die.

Would the absence of humans in depictions of ecosystems be because human beings play no role in them? Not likely. Levine (1996, 137) in describing an upland uncut forest in New Jersey concludes: "Ecologically, it turns out, it was a *scientific* mistake to treat humans as though they were not part of nature; to save woods, we need humans to burn them. And mistakes like that, if we extrapolate to, say, the rain forests of South America, or even the timberlands of the Northwest, have profound moral and political implications."

In short, we need to be able to ask questions like why a beehive is a part of nature but an apartment building is not. These assumptions, rarely stated explicitly, matter.

Perspective Constrains Research Orientations

Evelyn Fox Keller's book *A Feeling for the Organism* (1984) describes Barbara McClintock's Nobel prize-winning research program, including her research strategies and orientation. Although sometimes scientists are encouraged to adopt a disinterested, objective perspective, McClintock routinely followed the opposite strategy of attempting to place herself into the organism's point of view, in effect to become the organism. McClintock credits this strategy for much of her success.

Values Determine Research Methods

There are lots of animals in research labs in the United States and around the world. With few exceptions the only research participants that are free to come and go are human beings. Both humans and nonhuman animals are protected by ethical guidelines for research and institutional research boards, but these guidelines are different for humans and nonhuman animals. Thousands of rabbits have undergone pain and discomfort in the service of developing cosmetics that are safe for women and men (a nod to "guyliner") to wear.

We are making the obvious point that the kinds of methods employed in research reflect and are constrained by both societal and individual values. For example, Flo Gardipee, a Native American graduate student, was interested in the genetics of buffalos but felt strongly that extant methods for gathering DNA samples were intrusive and disrespectful to bison. She found a creative solution by developing a method for extracting DNA from buffalo feces (descriptions of her studies have used the charming title "What's the Scoop on Buffalo Poop?"). She has discovered that herds in Yellowstone Park are from two distinct genetic lines (Gardipee et al. 2007). In this case values led directly to innovation.

Values and Power Relations Determine Which Study Groups Are Privileged

It is no accident that research on birth control methods has largely focused on females, with the consequence that women and not men face the risk of side effects from birth control pills. For other health issues such as the treatment of disease there have been complaints that research has focused disproportionately on (white) males.

In the field of genetics studies Troy Duster (1996, 122) comments:

To illustrate the bearing of the perspective upon a "genetic prism," an investigation into the social backgrounds of the respective adversaries of the genetics and I.Q. controversy revealed that those scientists who pursued research claiming that high I.Q. was genetic and was located in privileged social strata were more likely to have come from upper-middle-class backgrounds; in contrast, those pursuing the environmental argument were more likely to have come from lower class origins.

Research Practices Are Constrained by Power Relations, Funding Priorities, and (Sometimes) Public Policy Debate

Although many biographies of scientists give the impression of brilliant individuals operating in isolation, current research is much more social in almost every sense of the word. In the "hard sciences" in particular the relevant unit is a multiperson lab, and many researchers develop from "bench scientists" into laboratory heads whose role focuses heavily on administration and management. These labs are expensive to run and for many scientists a primary activity is writing grant proposals. Inevitably, researchers face a compromise between what they would like to be doing and what they need to do to obtain research funds.

Arguably this state of affairs is positive because it allows national needs and policy concerns to influence the research that gets done. Although many scientists (and laypeople) were not pleased with President George

W. Bush limiting stem cell research, his actions do serve as a reminder that science alone does not dictate what is ethical. Issues such as whether or not it is ethical to clone human beings presumably should be driven by a wide policy debate—not some after-the-fact scrambling after a lab announces that it has successfully cloned a human.

Psychological Distance Affects Grain Size in Data Analysis

Facts are not just facts; they reflect a point of view. Here's a fact. The suicide rate among First Nations people in Canada in general and British Columbia in particular is about five times as great as the suicide rate for nonaboriginal Canadians. This is a serious problem and it calls out for interventions aimed at reducing the number of suicides. One could point to high rates of poverty, unemployment, broken families, and similar factors in looking for root causes. One could argue that intergenerational post-traumatic stress induced by years of federal policies has created these problems.

The implications of these grim statistics depend very much on the perspective and grain size of analysis. Chandler and LaLonde (1998) separately examined suicides rates by band ("band" is analogous to tribe, though a tribe may have more than one band, and has an associated governing body) in British Columbia. The first striking result was that the suicide rate varied dramatically by band and about 40 percent of bands had suicide rates of zero, far below the national average. The second result was that standard demographic measures like income, urban versus rural, and so on did *not* predict suicide rates. Chandler and LaLonde then developed a scale based on the number of different ways a given band was attempting to maintain or obtain their sovereignty (e.g., assert claims to traditional lands, language preservation efforts, and the like) and correlated this scale with suicide rates. They found a strong, systematic relationship between greater sovereignty efforts and lower suicide rates.

Note the change of perspectives associated with the different grain sizes. The overall statistics invite a deficit model where the natural intervention would be to target individuals who are "at risk." The analysis at the level of bands and resilience invites a focus more on bands than individuals and a focus on supporting sovereignty efforts. Both the province-level and band-level statistics are facts, but they carry vary different meanings. Chandler and LaLonde had worked with various bands in British Columbia for years, so it is likely that they didn't come on their insight by accident.

Grain size can include different levels of analysis, levels that vary in how they direct the attention of scientists (and policymakers). In the case of applying a genetics framework to health and other social problems Abby Lippman (1991) has raised the following concern: "For society, genetic approaches to health problems are fundamentally expensive, individualized, and private. Giving them priority diminishes incentives to challenge the existing system that creates illness no less than do genes. With prenatal screening and testing in particular, the genetic approach seems to provide a 'quick fix' to what is posed as a biological problem . . . leaving the conditions that create social disadvantage or handicap . . . largely unchallenged" (47).

Measures Are Affected by Values

Here we provide a single example recounted by Longino (2002, 12–13). She describes a controversy centering on how to measure the effects of plutonium in the air on lung tissue. This question involved assumptions about the distribution of plutonium in the lung. One approach assumed whole lungs were the relevant unit of analysis, but another suggested that there is a nonuniform distribution of inhaled plutonium in the lung and that some spots in the lung might receive doses much higher than average. If calculation of cancer risk is based on radiation dose, the stated and presumed risk will vary tremendously, depending on whether the whole lung or potential "hot spots" at a finer level of lung structure are taken as the relevant measure.

The standards set by the International Commission on Radiological Protection (ICRP) and the U.S. Atomic Energy Commission (AEC) were based on the whole-lung model. Many scientists argued that, in estimating hazards to humans and their environments and in the absence of firm evidence favoring one or the other, the safest course was to proceed on the assumption of greater rather than lesser harm; hence, the relevant unit of analysis should be spots within lungs, not whole lungs.

Values Embedded in Scientific Research: Summary

These are just a sample of the many ways in which values may affect how science gets done, even while the standards for scientific rigor remain constant and strong. It would be misleading, however, to assume that values are only in play for the context of discovery, leaving the context of justification/verification pure. Schwartz (1997) argues that discovery and verification cannot be conceptually isolated, at least in the social sciences, because discovery can lead to the creation of environments that provide a form of self-fulfilling prophecy that feeds back as a form of verification. One of his examples concerns racial differences in intellectual performance, where it has been established that stereotypes about blacks performing less well on intelligence tests contribute to bringing about this very effect (Steele and Aronson 1995).[1] Similarly one could readily imagine that in the absence of the Chandler and LaLonde (1998) analysis of suicide rates in individual First Nation bands, mediation programs might have been put into play that would work to solidify a deficit model of band mental health. Yet another example involves the study of "theory of mind," which consists of knowledge about other people's beliefs, desires, and intentions, including the idea that another person may have a false belief. Theory of mind has been extensively studied by developmental psychologists (e.g., Wellman, Cross, and Watson 2001), and it is tempting to conceptualize it as a universal endpoint of typical conceptual development. Recently, however, a number of researchers, mostly anthropologists (see Luhrmann 2011 and commentaries), have suggested that cultural notions about minds affect the very nature of this endpoint and undermine the assumption of a universal endpoint. Finally, we end this chapter with an example of pernicious niche construction that suggests why women may not be "well suited" for the field of philosophy.

The Cost of Discouraging Diversity of Perspectives

We owe this analysis to Buckwalter and Stich (2010), and it concerns the status of intuitions in philosophy. Intuitions are of central importance in philosophy because a great deal rides on the construction of scenarios and dilemmas that compellingly demonstrate (that is, call on our intuitive understanding of) some thesis or phenomenon. We should note that recently philosophy has taken an empirical bent and has systematically examined whether intuitions are indeed shared. Sometimes the assumed consensus shows variation as a function of social class, gender, or number of philosophy courses taken.

Buckwalter and Stich (2010) review a number of studies demonstrating gender differences in intuitions concerning a range of philosophical questions. This is all well and good and of interest to both philosophy and psychology. But there's a catch. Almost 80 percent of the faculty in philosophy departments are males, whose intuitions, as we have just seen, may depart from those of female graduate students. Buckwalter and Stich suggest that this gender disparity tends to perpetuate itself by what they call a *selection effect*. In their words:

Students come to philosophy with somewhat different intuitions about many standard philosophical thought experiments, and as we have shown, in many cases there are statistically significant differences between women's intuitions and men's. However, most of the faculty members who get to say which intuitions are correct (and "obvious") are now, and always have been, men. So women students are more likely than men students to find that their intuitions about the thought experiments discussed in their philosophy classes are at odds with those of their instructor. If it is indeed the case that students (of either gender) are less likely to continue in philosophy if their intuitions do not accord with those of their instructor, then all the elements of a powerful and cumulative selection effect are in place—a selection effect which "filters out" a greater proportion of women than of men. (30)

Later on they add:

As a result, the men will tend to get better grades and be more inclined to continue in philosophy, while the women will get poorer grades and be more inclined to look elsewhere. If this story, or even a substantial part of it, is more or less on the right track, then difference in philosophical intuition is a crucial factor in unleashing a cascade of events that increasingly skews the gender distribution in philosophy courses toward men and away from women. (33)

We think that Buckwalter and Stich are right to be worried about a selection effect and that selection effects are widespread. On our account, selection effects include not just intuitions but also a wide range of implicit and explicit practices that may mediate engagement with many fields of learning, including science. If white male practices dominate science or middle-class practices dominate public education, then those who do not fall into those categories may never have their own practices recognized or valued and may be placed at risk for disidentification and alienation.[2]

Summary

Differences in values provide different perspectives on nature, each of which may yield useful insights. The next step in our argument is to note that values vary as a function of factors such as gender, social class, and culture.

From these points it follows that diversity in researchers and research perspectives makes for better science. Chapter 5 provides further examples of how diversity of values and perspectives influences theory and practice in science.

We close with a quote from Levins and Lewontin (1985, 5): "To do science is to be a social actor engaged, whether one likes it or not, in political activity. The denial of the interpenetration of the scientific and the social is itself a political act, giving support to social structures that hide behind scientific objectivity to perpetuate dependency, exploitation, racism, elitism, colonialism."

5   Science Reflects Who Does It

The males of certain hymenopterous insects [bees, wasps, ants] have been frequently seen by that imitable observer, M. Fabre, fighting for a particular female who sits by, an apparently unconcerned beholder of the struggle, and then retires with the conqueror. (Darwin 1872/1998, 69)

This sentence from Darwin reflects a passive view of female bees, ants, and wasps. Would a Charlotte Darwin have shared this view or might she have discovered a greater role for the female of the species? Obviously one can't say either way, but one might have some suspicions. In this chapter we will describe studies that bear on individual and team researcher diversity that strongly suggest that the way science gets done is not independent of who's asking the questions. We continue by examining the role of gender in science and the idea that there may be a distinct feminist science. The subsequent step will be to turn to a consideration of how culture may shape science.

Is Science Androcentric?

One notion that appears to be well supported is the claim by feminist scholars that male-dominated science has been associated with an implicit androcentric focus. For example, in describing the lack of research on the role of language in evolution Ruth Hubbard (1979, 65) laments: "It is likely that the evolution of speech has been one of the most powerful forces directing our biological, cultural and social evolution, and it is surprising that its significance has been largely ignored by biologists. But, of course, it does not fit with the androcentric paradigm. No one has ever claimed that women cannot talk; so if men are the vanguard of evolution, humans must have evolved through stereotypically male behaviors of competition, tool use and hunting."

Similarly, the very (androcentric) terminology used to describe some phenomenon may constrain or bias the focus of research. For example, in discussing reproductive behaviors, it likely makes a difference whether one talks about a passive "female receptivity" or a more active "female choice."

And while we're on this topic, a focus on reproductive success as a driver of evolution may lead to a view of sex that is bound to reproduction. For example, Lloyd (1993) describes a bias to think of every aspect of female sexuality as concerned with reproductive functions, and she notes that "if estrus is defined for females as willingness to engage in sex then bonobos are in estrus 57 to 86% of the time!" Not very likely. Lloyd also mentions that peak human female desire is right before or right after menstruation, when women are infertile. So maybe sex doesn't just have a single function.

Longino (1990) argues that male and female stereotypes pervade research on gender and child development, most notably in the construct of describing some girls as "tomboys." She notes that the use of this term accepts the cultural assumption that lively activity is for boys and quiet, role-based play is for girls. Furthermore, she suggests that this bias may influence the selection of data, highlighting the presence or absence of some behavioral factors and overlooking or downplaying others.

The same gender stereotypes hold for studies of parenting. As one feminist psychologist argued: "It is scientifically unacceptable to advocate the natural superiority of women as child-rearers and socializers of children when there have been so few studies of the effects of male-infant or father-infant interaction on the subsequent development of the child" (Wortis 1971).

An androcentric bias almost reaches the level of comedy when we shift attention to the level of egg and sperm and associated stereotypes. We well remember being taught about the egg, more or less just sitting there, while millions of sperm, like Olympic swimmers, race to fertilize the egg and there is only one winner. (We also remember reading about how menstruation expresses "nature's disappointment," but we never saw anything about the millions [or millions minus one] of sperm who could have stayed home.)

As Emily Martin (1996) laments, this view nicely recapitulates stereotypic male and female roles, and she critiques it as a misleading model of fertilization. What if the egg exercises some selection over the sperm? We probably would not notice any role for the egg if we have already bought into this framework and its implicit assumption of a passive egg. However, Martin cites research (Baltz, Katz, and Cone 1988) showing that the egg surface must be designed to trap the sperm which otherwise move back and forth, flopping like fish on land, rather than going forward. Does that design select for some sperm rather than others? If we don't look, we won't see, so metaphors matter.

An androcentric orientation might extend beyond the focus of attention to include the kinds of explanatory mechanisms that tend to come to mind. We noted before that (male) evolutionary theorists have been criticized for their focus on competition in a hostile world. Although this orientation may be useful for many purposes, it tends to neglect opportunities for cooperative solutions. Bazerman and Neale (1992) have observed that people often have a zero-sum orientation (if you win, I lose and if I win, you lose) that blinds them to integrative win-win opportunities.

Bob Axelrod (1987, 1997) gained a great deal of well-deserved attention for designing the winning system for a game theory tournament where researchers entered their strategies for playing variations on the Prisoner's Dilemma (PD) game. In the classic form, if one prisoner defects (confesses) while the other cooperates (doesn't confess), the defector gets a plea bargain and a light sentence and the cooperator gets a harsh sentence. If they both confess, they both get a sentence length between the light and the harsh alternatives. If neither confesses, then they both get off. Game theory dictates that one should minimize losses by confessing, but obviously if the two prisoners trust each other and cooperate they will be better off.

Different instantiations of the game have slightly different payoff structures; for present purposes we will assign points to the various outcomes as follows: If both players cooperate, then they both gain (say, three points). If both defect, they both lose (say, one point). But if one cooperates while the other defects, then the cooperator loses (two points) and the defector gains (four points). So cooperating is a very good strategy as long as the other player cooperates, but defecting can also pay off if the other player cooperates (is a sucker). For the tournament in question, each system or strategy acted as a player; the players were randomly paired with other players for many iterations.

There were quite a few entries, many of which had been programmed to have very complex strategies, but Axelrod's algorithm was a simple "tit for tat." It would start out cooperating, and if the other player defected on a round, it would defect on the next round; if the other player then cooperated, it would cooperate on the next round. One of the reasons Axelrod's tit-for-tat strategy drew so much attention is that formal game theory and many of the other entries had a much more competitive bent. This is a long way for us to make the point that the best way to compete may not be to compete.

With respect to androcentric bias we cannot give a better summary than that of Keller and Longino (1996, 30): "It is not true that 'the conclusions of natural science are true, and necessary, and the judgment of man has nothing to do with them,' it is the judgment of women that they have nothing to do with." The dominance of men in science does not mean that an androcentric science is bad science, just limited science.

Feminist Science?

It is hard to imagine a National Academy of Female Sciences and a National Academy of Male Sciences rather than a single academy. However, given the background we have covered it is easy to see that female scientists might bring somewhat different perspectives and values to the table.

We're not proposing to review the vast literature on feminist science; toward that purpose, see scholars such as Ruth Bleier (1984), Donna Haraway (1988), Helen Longino (2002), Sandra Harding (2006), and Evelyn Fox Keller (1986, 2010), and edited volumes by Gowaty (1997) and Strum and Fedigan (2000). Instead we'll limit ourselves to offering a few reasons for expecting feminist science to be different, followed by a few cautions about our claims.

1. Feminist science is less likely to be androcentric. We qualify this suggestion by saying "less likely" because women have to compete in a male-dominated science and success in the field may depend, in part, on adopting the (androcentric) practices and orientations that the field expects to see.

2. Feminist science may be more likely to attend to context and interactions, reflecting a relational focus.

3. Feminist science may differ from male-dominated science to the extent that women have different values than men. In chapter 4 we pointed to a range of ways in which values may affect scientific practices.

4. Feminist science may differ from male-dominated science because power relations (domination) differ. We do not develop this point here other than to note that there is evidence that a position of power may be associated with less effective perspective taking and inattention to context (Galinsky et al. 2006).

5. Feminist science may differ from male-dominated science because the experiences of women differ from those of men. This thesis about diversity of experiences applies in general, but it certainly holds for the case of gender.

Now for a few disclaimers. First, it is not our intention to essentialize the relation between gender and the ways and whys of doing science. We make no claims about any biological hardwiring ("women's nature") that, for example, would make attention to context and relations inaccessible to men or a focus on abstract linear causal models inaccessible to women. Nor do we claim that social factors, stereotypes, and the like associated with one culture are going to generalize across all cultures or even hold uniformly within a single culture. Our claim is this simple: given a three-way option of selecting among equally qualified researchers either (1) twenty males, (2) twenty females, or (3) ten males and ten females to work on some problem of national interest, we would take the third option (see Page 2007 for evidence that diversity can even trump ability). We turn now to two detailed examples of researcher orientation effects, one starting with gender and then shifting to culture and the other focused on culture.

Gender, Culture, and Morality

Initially, the study of morality in psychology was primarily an examination of how children form moral concepts.[1] Jean Piaget is often credited with introducing the concept of moral development in psychology (Piaget 1932/1997). He described several stages children traverse in order to form a coherent moral concept, from a self-centered to a principle-based morality.

Inspired by the Piagetian approach, Kohlberg (1973, 1984) devised a six-stage model of moral development progressing from the most relativistic view of morality to the most universalistic, presumably the most advanced form of moral reasoning. In the first level of the model, people behave morally by following the constraints of some extrinsic set of rules (e.g., to avoid punishment). In the intermediate stages, people uphold the moral values of the social system and fulfill interpersonal obligations as mandated by the community. The final stage, "Universal Ethical Principles," consists of the realization that ethical principles are universal and independent of social systems.

Enter Gender

Although Kohlberg's model received considerable support and was widely influential, Carol Gilligan argued that the model was male-oriented and failed to capture gender differences in moral reasoning. In her revised model, Gilligan claimed that women possess certain unique moral principles and modes of reasoning leading women to have different perspectives on ethical dilemmas than men (Gilligan 1982). According to Gilligan, males may evaluate themselves and others around them on the basis of abstract principles such as justice or equality, but women measure themselves in terms of particular instances of care. Women tend to have a more contextualized view of morality, one that takes into account relationships between people rather than a notion of an overarching moral principle that applies uniformly across situations.

As a hypothetical example, imagine a young boy, Jake, being interviewed about the Heinz dilemma, a frequently used example from Kohlberg's interviews. It goes like this:

A woman was near death from a special kind of cancer. There was one drug that the doctors thought might save her. It was a form of radium that a druggist in the same town had recently discovered. The drug was expensive to make, but the druggist was charging ten times what the drug cost. He paid $200 for the radium and charged $2,000 for a small dose of the drug. The sick woman's husband, Heinz, went to everyone he knew to borrow the money, but he could only get together about $1,000 which is half of what it cost. He told the druggist that his wife was dying and asked him to sell it cheaper or let him pay later. But the druggist said: "No, I discovered the drug and I'm going to make money from it." So Heinz got desperate and broke into the man's store to steal the drug for his wife.

After being presented with this scenario participants are asked, "Should Heinz have broken into the laboratory to steal the drug for his wife? Why or why not?" (Kohlberg 1973, 638)

Suppose Jake is certain that Heinz should steal the drug to save his wife's life, because the value of life supersedes all else. A young girl, Amy, in contrast, may cite the importance of the wife's survival to the husband's well-being, and acknowledge that in a different context she might have a different opinion.

A Kohlbergian-model scoring system would relegate Amy to a conventional stage of moral reasoning that highlights interpersonal relationships, but arguably Amy's moral reasoning is more complex than Jake's. Unlike Jake, she reasons broadly that in such a scenario there is no "right" answer. Gilligan suggests that even at equal stages of moral maturity by Kohlberg's standards, men and women may provide very different justifications for their judgments. Gilligan's observations suggest that Kohlberg's model is incomplete because it only approaches morality from an ethics of justice.

Had the modern study of moral reasoning started with Gilligan and were the field dominated by women, then it may well be that male researchers would come upon the scene and point to limitations in Gilligan's model. So one reason diversity may matter is that it is associated with different senses of how things are.

Enter Culture

Snarey (1985) employed a line of argument similar to Gilligan's to critique the Kohlbergian framework. He reviewed findings testing Kohlberg's model from twenty-seven different cultural groups, including small-scale societies and involving a wide array of religious beliefs. By this measure, the final, mature stage of moral development was absent in all of the eight traditional tribal or village folk societies. Instead of endorsing a deficit model, Snarey concluded that Kohlberg's model incorporates cultural bias, because it had been developed largely with urban populations. Had Kohlberg's research originated in a small-scale society, his theory of moral development might have been very different, Snarey suggests.

Snarey also noted that Kohlberg's model does not account for moral principles engendered by Eastern religious traditions. For example he cites a study by Gielen and Kelly (1983) conducted with Buddhist monks from Ladakh, a Tibetan culture in northern India, which produced the counterintuitive result that, by the Kohlbergian system, monks would receive lower moral reasoning scores than laypeople. Gielen and Kelly concluded that Kohlberg's model was insufficient for understanding the principles of cooperation and nonviolence.

Another set of researchers found that the scores of Taiwanese participants could change depending on whether they were taking a subordinate role (son) or a superordinate role (father) while reasoning about a dilemma (Lei and Cheng 1984). Role-dependent judgment also is outside the scope of Kohlberg's theory. Overall, these studies show that conclusions from studies of moral cognition depend very much on who's asking the questions. We turn now to another illustrative example of how scientist culture affects science, in this case focusing on primatology.

Culture and Science: A Mini Case Study

We don't want to stray too far from our overall aims, but the ways in which primate behavior has been studied in India, Japan, and the United States make for intriguing contrasts—contrasts that reflect different environmental contexts and cultural values. (Coincidentally, for a fascinating analysis of the role of scientist gender in primatological research see Donna Haraway's 1989 *Primate Visions*.)

First of all, U.S. primatologists historically adopt the stance of being "minimally intrusive." The scientist observer is "outside the system" and records behavior at a distance. (If you're a typical reader this statement may well puzzle you, as images of Jane Goodall and close relationships with apes immediately come to mind. But we're talking about a research tradition that started in the 1930s and 1940s, one initially dominated by male researchers. Times have changed.) In India people and other primates coexist, and Indian primatologists would find it odd to impose a distance between observer and observed or to conceptualize human beings as outside the life space of their fellow primates. Although monkeys and humans do not live together in Japan, Japanese primatologists find it reasonable to feed the objects of study (provisioning also makes observation easier) and to develop a relationship with them. Americans would tend to see feeding monkeys as an intrusion and as contaminating "naturalistic observation."

Given these differences in background assumptions, one might also expect to see differences in other scientific practices. In the following we focus on the contrast between U.S. (and European) and Japanese primatology, drawing heavily on analyses by Pamela Asquith (1996) and Hiroyuki Takasaki (2000).

Japanese, European, and U.S. studies of primate behavior began within a decade of one other in the 1930s and 1940s. European studies of primate behavior go at least back to Sir Solly Zuckerman's 1932 *The Social Life of*

Monkeys and Apes, which was considered a milestone in primatology. It was based on observing captive baboons, and it was not recognized at the time that the behavior displayed by the overcrowded hamadryas baboons was highly atypical and reflected the captive conditions.

Western primatologists typically adopted a biological perspective, and early studies tended to focus on male dominance and the associated mating access. Other social behaviors were also viewed from this male-ranking lens. The implicit assumption was that researchers were studying species-specific behaviors and that contexts or settings were not important. There was little attention to individuals qua individuals except to trace dominance hierarchies, and rarely were individuals or groups tracked across extended periods of time.

Japanese researchers had a distinctly different orientation, according to Asquith's review and analysis. They tended to adopt a more cultural perspective and sought phenomena among primates parallel to what occurs in Japanese society. Since status and social relationships are given much attention in Japan, it was natural to look for parallel phenomena in the monkey groups. An important aim was to understand what position each individual held in their society. Therefore it was natural to pay close attention to individuals and to track the history of both the individual and the group across time. As a result, Japanese reports of behavior were more personal, anthropomorphic, and richly detailed as to individual animals' life histories. These observations had implications for both method and theory that were overlooked in the West, despite the fact that Japanese primatologists were publishing in English-language journals.

These differences in orientation were associated with striking differences in the insights that the various research programs generated. For example, Japanese primatologists discovered that male rank was only one factor determining social relationships and group composition. They found that females had a rank order too, and that the stable core of the group was made up of lineages of related females, not males (Asquith 1996).

The longer-term studies of Japanese researchers also allowed them to notice that maintaining one's rank as the alpha male was not solely dependent on strength (Itani 1961). For example, in describing Jupiter, the alpha male of the Takasakiyama troop, Itani (424) says, "He could hardly have maintained his position without something other than the strength of his teeth and arms. Our numerous records convinced us that it had some connection with the individual's influence and achievements, and particularly with the confidence of the females who make up the center of the troop."

Japanese primatologists also discovered within-species differences in social organization. In the West, until Thelma Rowell's (1966) observation of differences among groups of forest- and savannah-dwelling baboons of the same species, it was thought that you could identify general species-specific behavior by observing a single group.

As we noted earlier, Japanese primatologists often did not hesitate to offer primates food if it would help their observations. The Japanese practice of providing food to animals also led to important insights. One breakthrough came when researchers lured monkeys to a sandy beach with sweet potatoes. This allowed the researchers an unobstructed view of the monkey troop. This also led to the serendipitous finding of "cultural transmission." One monkey hit on the idea of washing a sweet potato to get rid of the gritty sand on the its skin and other monkeys noticed and imitated this behavior (Itani 1958). Researchers were then able to trace the transmission of this cultural behavior across generations (Kawai 1965).

With their long-term studies, and willingness to attribute more complex motives to animal behavior, Japanese scientists discovered much about primates that Westerners recognized only considerably later. Western primatologists tended to dismiss Japanese observations as subjective and anthropomorphic (Frisch 1963), but ultimately conceded the accuracy of their findings. Asquith suggests that the Western attitude was that the Japanese find the right answer in the "wrong manner." But arguably it is precisely the assumption that monkey social organization parallels Japanese society that allowed them to make their breakthrough observations, and it may be that the more "rigorous" Western methodology prevented Westerners from doing so. The "less rigorous" approach in Japanese primate studies was actually a carefully considered methodology.

With respect to differing approaches to scientific observation, Takasaki (2000) sums up the literature we have just been discussing this way (164):

Although science is supposed to be universal, each scientist is a cultural being, affected by his or her cultural background. . . . Intercultural and school differences, as well as individual idiosyncrasies are essential to the advancement of science. Such differences are comparable to the biological diversity and gene stock that are necessary for the global ecosystem's stability and for evolution of new diversity. There are always some parts of the world that may be viewed better upside down.

A Comment on Japanese Publishing Style

Asquith (1996, 255) suggests that for the Japanese,

First, their focus in certain fields and at certain times in the history of a discipline may be directed more toward a building up of details to support one position or another. This may be seen as necessary before stating any specific position. What may have contributed to the notion that Japanese scientists tend to build on others' theories, rather than develop their own, is the way in which they write their reports. Scientific reports written in Japanese do not typically state conclusions (see Leggett, 1966; Motokawa, 1989). Instead, they try to describe one fact from various points of view. These points may be connected by imagery to other points, rather than by strict logic. Why do Japanese scientists not state a firm conclusion? It would, they say, close their world.[2]

Sociologists of science have written about a number of other cultural differences in how scientific research is organized and conducted. For example, in the U.S. Office of Naval Research, the Army and the Air Force are major funders of research, including social science research, but in Japan researchers tend to avoid military funding, perhaps as an outgrowth of post–World War II controls. Another obvious difference is in how labs are structured. For example, in Europe laboratories tend to have a stronger hierarchical structure than in the United States. These sorts of differences can affect what research gets done as well as how it is done.

Another Look at Views of Nature

Another source of variation in science practices that we have mentioned earlier and will take up in detail later on is how human beings are conceptualized in relation to the rest of nature. Some cultures may see human beings as more distinct from the rest of nature than others do. For example, in English the term "animal" is ambiguous and sometimes is used in a sense that includes humans but sometimes is used contrastively with human. Indonesian, however, only has the contrastive sense of "animal." Although it may seem minor, this difference is associated, at a minimum, with cultural differences in children's inductive reasoning about biological kinds (Anggoro, Medin, and Waxman 2010; Anggoro, Waxman, and Medin 2008).

Keller (1992) argues that it is impossible for scientists to describe nature in a way that is not mediated by language (including the language of mathematics). She says (30): "Even in the loosest (most purely descriptive) sense of the term *law*, the kinds of order in nature that laws can accommodate are restricted to those that can be expressed by the language in which the laws of nature are codified."

Why is this important? Because there is no "neutral language" that does not take a perspective. For example, the field of psychology is full of descriptive language like "executive function," "undermining intrinsic motivation," and "fundamental attribution error" that not only describe phenomena but also tend to reify them. Often these phenomena arise from some paradigm or procedure designed to demonstrate them unambiguously. But when these phenomena are given a label, it becomes tempting to generalize them to all situations where the label applies.

Let's be concrete. In a clever series of studies, Lepper, Greene, and Nisbett (1973) demonstrated that external rewards may undermine intrinsic motivation. So, for example, they might provide children with a reward for playing with a toy and find that children who were rewarded actually played with the toy for less time than children in a control condition who were not rewarded. These are important findings, and understanding when they do and do not appear is an important theoretical and empirical challenge. Note, however, that, given the handy summary that "external rewards undermine intrinsic motivation," it is all too easy to assume that external rewards compete with self-motivation wherever they appear. But to our knowledge, no winners of the Nobel Prize have turned it down on the grounds that it would ruin their interest in their respective fields. So clearly the original results for Lepper and colleagues have some generality but they cannot be extended based solely on the language used to describe them, because that implies universality not in evidence (for relevant critiques and reviews, see Eisenberger and Cameron 1996; Ryan and Deci 2000; Lindenberg 2001; Cameron 2006).

A special case of language that pervades life in general (Lakoff and Johnson 1980) and science in particular is the use of metaphors. Metaphors and analogies can be very productive for researchers (e.g., Gentner and Markman 1997), but the phenomena themselves may be understood differently based on what they are compared with. To give a perceptual example, the drawing in the middle in figure 5.1 generally is described as three-dimensional when it is compared with the figure on the left, but as two-dimensional when compared with the figure on the right (from Medin, Goldstone, and Gentner 1993). These drawings may be a trivial example, but metaphors for nature likely are not. Western science has been pervaded by the cultural notion that nature is female and that scientists labor in hopes of getting her to reveal "her secrets." Some go so far as to suggest that this metaphor includes the idea that nature must be grasped, dominated, and, in short, "conquered."

Of course, linguistic differences are just one component of a complex pattern of orientations toward nature. Stephen Kellert and his associates (e.g., Kahn and Kellert 2002; Kellert 1996, 1997) have conducted cross-cultural (and developmental) studies of how people value nature and have constructed a typology featuring nine distinct types of orientations toward nature, ranging from dominionistic focusing on the mastery and control of nature, to utilitarian focusing on value to humans, to aesthetic and humanistic focusing on the beauty of nature and emotional bonding with nature, respectively. The case of Flo Gardipee's studies of buffalo (buffalo poop), recounted in chapter 4, reflects different values and associated practices than a dominionistic or utilitarian orientation would dictate. Later on we will adopt modifications to Kellert's scheme in discussing our own research.

Science, Culture, and Policy

This is a good place to comment on the policy side of things. Just as patients may tend to seek the advice of doctors, policymakers often need scientific expertise to determine what levels of mercury or bacteria in the drinking water are safe (or acceptable risks), how to measure air quality, whether genetically modified crops pose a hazard, and how serious global climate change is. Although there is a very strong consensus on the reality of global climate change and although no one thinks that mercury does not affect human health, it nonetheless is still the case that the culture of the scientist

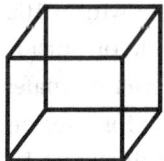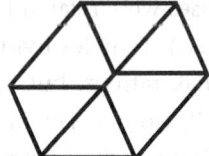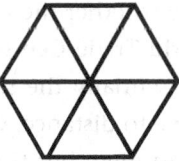

Figure 5.1
Example of how comparison affects perception. When the middle figure is compared with the left figure, the middle figure tends to be described as three-dimensional, but when compared with the figure on the right it tends to be described as two-dimensional.

matters. In the final section of this chapter we describe what we call the "white male effect."

Let's start with a couple of intriguing observations. Finucane et al. (2000; see also Kahan et al. 2007) conducted an extensive national survey of people's perceptions of different forms of risk. They found that people generally agreed on the relative seriousness of risks but that one subgroup, white males, tended to give systematically lower estimates of risk. The "white male effect" appeared to be caused by about 30 percent of the white male sample who judged risks to be extremely low. The remaining white males were not much different from the other subgroups with regard to perceived risk.

The low-risk-perception subgroup of white males also held very different attitudes from those of the other respondents—they were more likely than the others to: (1) agree that future generations can take care of themselves when facing risks imposed on them from today's technologies (64.2 percent vs. 46.9 percent); (2) agree that if a risk is very small it is okay for society to impose that risk on individuals without their consent (31.7 percent vs. 20.8 percent); (3) agree that science can settle differences of opinion about the risks of nuclear power (61.8 percent vs. 50.4 percent) and that government and industry can be trusted with making the proper decisions to manage the risks from technology (48.0 percent vs. 31.1 percent); and (4) agree that we can trust the experts and engineers who build, operate, and regulate nuclear power plants (62.6 percent vs. 39.7 percent) and disagree that local residents should have the authority to close a nuclear power plant if they think it is not run properly (50.4 percent vs. 25.1 percent).

Fischhoff et al. (2003) did a national survey of perceived risks of terror-related events as well as more routine risks (e.g., being the victim of other violent crime) over the course of one year. Given the events of 9/11 they also looked to see whether the responses were related to living within 100 miles of the World Trade Center (WTC). Females living inside or outside this zone gave essentially the same risk ratings, but judgments by males were very sensitive to distance. Overall, women saw more risk for a terror-related attack than men, but this difference disappeared within the WTC zone. Political party affiliation was also related to the distance effect. Distance was correlated with terror risk judgments for Republicans but not for Democrats or other non-Republicans. Adult male Republicans were more sensitive to distance than were all other adult males for terror risks but not for routine risks. The net effect of these factors is that the correlation between distance and fear of being hurt in a terror attack was quite large for white, male Republicans, suggesting that this group has a special view of risk (see also Vaughan 1993).

What holds for the general public appears to hold for trained scientists as well. Kraus, Malmfors, and Slovic (1992) found gender differences among expert toxicologists in risk assessments. For example, women toxicologists were more willing than men to say that if a chemical causes cancer in animals, it will likely cause cancer in humans. Higher risk ratings for standards was also associated with disagreeing that "economic growth is necessary for good quality of life" and disagreeing that technology is important for social well-being (though this study was done before the Facebook, Twitter, texting era). Similarly, Barke, Jenkins-Smith, and Slovic (1997) found that female physical scientists judge risks from nuclear technologies to be higher than do male physical scientists.

What drives this white male effect? There have been a number of suggestions ranging from political orientations to "worldviews" or "cultural models" such as individualism versus egalitarianism (see Dake 1992; Douglas and Wildavsky 1983; Gastil et al. 2011; Kahan, Jenkins-Smith, and Braman 2011; Lima and Castro 2005; Peters, Burraston, and Mertz 2004; Sjöberg 1998), but we're not going to examine them or try to choose sides.

The critical observation is that the white male effect appears and that it is evident for both laypersons and scientists. As Kahan et al. (2006) note, values determine what significance individuals attach to the risks and consequences of factors such as global warming, nuclear power, pollution, gun control, and even casual sex.

The same appears to hold in those cases in which the individuals in question are scientists. If risk assessments reflect (cultural) values of the scientists making them, then the public is largely being given white male values. Unless you are happy with the idea of a white male effect being hidden under the cloak of an objective, neutral science, it seems like a good idea to recognize that the answers to questions that science asks may depend in important ways on who's asking.

In the words of Slovic (1999, 689): "Whoever controls the definition of risk controls the rational solution to the problem at hand. . . . Defining risk is thus an exercise in power."

6   Culture and Issues in Cultural Research

What has been thought of as the mind is actually internalized culture. (Hall 1976, 168, italics in original)

Pretty much everyone thinks they know what culture is—the trouble is that people may not agree with each other. A few people think of culture as something other people have, something that explains the peculiar things they do or the funny ways they think (Ross 2004). They themselves see the world directly and objectively, but others have their perceptions biased by the lens of culture. More often people tend to think of culture in terms of shared histories, customs, and beliefs. Still others see culture as corresponding to "a people" and remain agnostic concerning the status of shared values. For now let's tentatively agree to think of culture (in a big-tent manner) as "a way of life."

Although none of these definitions of culture necessarily require comparison, often people know the cultural by noting cultural differences. That is, it may appear that there is a natural or normal way that things just are, and we are surprised that ways of life could be and are different in another culture. From what people eat (Rozin 2007; Rozin, Kurzer, and Cohen 2002) and marriage customs to sleeping arrangements (e.g., Shweder, Jensen, and Goldstein 1995), intriguing cultural differences are found. Medin's lab includes scholars from China, Bulgaria, Italy, India, and the United States, and not too long ago when several of them were chatting about culture, they discovered that their group of five had four distinct ways of tying their shoes (Iliev 2012).

The study of culture is as difficult as it is fascinating. Few research areas are as problematic or have such an appalling history as the study of culture. Cultural research has been used as an instrument of colonialism to justify subjugating people whose thought processes are "more primitive" and inferior in other ways (e.g., other cultures have also been seen as morally degenerate). In short it is easy to interpret cultural differences in terms of a deficit model, where the cultural group that differs from "us" (the cultural group in power) is seen as having failed, where "failed" means not performing in accordance with the privileged standard. For these reasons some have argued that the very construct of culture is so susceptible to stereotyping and essentializing other groups that it should be discarded altogether (see Borofsky et al. 2001; Brumann et al. 1999).

For convenience and to get our discussion going, we tentatively define culture as the knowledge, values, beliefs, and practices among a group of people, usually living in geographical proximity, who share a history, a language, and cultural identification. Importantly, we view knowledge, values, and beliefs as causally distributed patterns of mental representations, their public expressions, and the resultant behaviors and practices in given ecological contexts.

Sorry for the abstract language, but we need to at least point to relevant facets of culture. Cognitive psychologists tend to think of culture as strictly in people's heads and don't usually pay much attention to the environment, artifacts, or even other people. Conversely many anthropologists appear to equate culture with everything *but* what is in the minds of individuals. People's mental representations interact with other people's mental representations to the extent that those representations can be physically transmitted in public media (language, dance, signs, artifacts, etc.). These public representations, in turn, are sequenced and channeled by ecological features of the environment (including the social environment) that constrain interactions between individuals and between individuals and their surrounding environment. It is also important to emphasize that ideas, or mental representations of them, do not circulate in a vacuum—ideas are contextually embedded. This context may include framework theories that are so basic and so much part of our backgrounds that normally we are unaware of them.

It is important not to essentialize culture. While there are historical continuities that connect individuals across generations (Lee, Spencer, and Harpalani 2003), describing culture in categorical terms to distinguish groups of people often leads to statements that attempt to describe the "essence" of groups. This can lead to stereotyping, such as the idea that Asian children are good at math or that girls don't like science and aren't good at it. Such statements assume that all group members share the same set of experiences, skills, and interests (Gutiérrez and Rogoff 2003).

In our own research on culture we almost never are focused on what or how the average person of some cultural group thinks. Instead our aim is to identify different frameworks or ways of thinking that may be correlated with cultural memberships. In this sense we are more concerned with what cultures a framework has than what frameworks a culture has. On this view a given culture may provide more fertile ground for some sets of ideas than for other sets of ideas (and different cultures have different soil characteristics).

Culture is not a trait or homogeneous body of knowledge possessed by some community, but rather (highly variable) ways in which people live. One could say that people "live culturally," and in the previous chapters we have been arguing that scientists "do science culturally." Again we have to be careful because our intention is to convey not only that there may be distinct practices associated with (the field of) science, so science has a culture, but also that scientists' practices reflect their own culture and cultural values.

One reason to think carefully about culture is that definitions of culture affect how researchers go about studying culture. If the study of culture is conceptualized as identifying shared norms and values, it is natural to assume that individuals become part of a culture through a process of socialization. It also means that once you have identified a consensus on these norms and values, you don't need to keep asking about it, any more than if you ask five people what day it is and all five agree.

If instead culture is seen as dynamic, contested, and variably distributed within and across groups, it is natural to see cultural learning as involving a reciprocal relationship between individuals' goals, perspectives, abilities, and values and their environment (Hirschfeld 2002). On this view, socialization partially depends on agents or others who are caregivers as well as an individual's interpretation of and reaction to their environment. In addition, the task of a researcher goes beyond determining the consensus and may include tracking down within-culture sources of variability in values, practices, and the like.

At one point Medin liked the idea that cultures are like species (Atran, Medin, and Ross 2005). Dan Sperber had warned us against using this analogy on grounds that people tend to essentialize species so we would be inviting them to do the same for culture. Perhaps out of stubbornness, we stuck with the analogy. We suggested that just as it was difficult for biology to discard the essentialized notion of species in favor of species as a historical, logical individual (Ghiselin 1981), it is difficult to abandon the commonsense notion of culture as an essentialized body (of rules, norms, and practices). We know now that in biology, it makes no sense to talk about species as anything other than more or less regular patterns of variation among historically related individuals. Nor can one delimit species independently of other species. So, too, it makes little sense to study cultures apart from patterns of variation. Although Medin continues to agree with this last point, Dan was right.

Not having learned our lesson that metaphors are often misleading, we now propose the following alternative analogy: *Cultures are like ecosystems.* For openers it gets away from the tendency to essentialize cultures. Second, it invites one to pay attention to within-culture relationships. Third, it raises some challenging questions concerning how to conduct cross-cultural comparisons. And last but not least, it encourages attention to system-level dynamics rather than focusing on some component in isolation. The reader will readily grasp the issues at stake from the following tongue-in-cheek story.

The Tortilla Story

Suppose that we are researchers from Mexico trying to understand the psychology of food in Mexico and we see tortillas as a fundamental part of this psychology. We decide to do some cross-cultural research where we assume that some analogous food is central in each culture.

One team goes to Ethiopia and discovers injera, the flat round bread used daily as a base for meals. Another goes to India and examines different forms of naan. In Italy, we discover pizza. It is starting to look like we have a cross-cultural universal going in terms of round flat cooked grains upon which other foods are placed. Visiting some Native American communities, we might find that frybread is analogous to the tortilla. Finally, we go to U.S. middle-class communities and, after some considerable research, we decide that perhaps pancakes are the best match to tortillas.

But the closer we look, the less happy we become. Our first embarrassment is that when we do back-translation from English to Spanish for pancake we do not obtain a quick answer, and when the answer comes, it is sopapilla, not tortilla. When we go on line to check Spanish-to-English dictionaries for "tortilla" we get "tortilla," and for "sopapilla" we get "sopapilla," apparently untranslatable constructs. Later on when we add an Italian researcher to our team, she makes us nervous by politely suggesting that pasta is a better match for tortilla than pizza is.

This trend is not good for our morale. When we inquire further into naan and its significance in India, our expert informant says:

Well, the funny thing is that people don't usually make naan at home. It requires a tandoor (a type of oven) which most people don't really have, so if you want naan, you usually go to a neighborhood bakery. The traditional flatbread that everybody eats at home is called roti (at least in the north, while in the south rice is the preferred source of carbs). And it was quite the necessity for women traditionally to be good at making rotis—with good encompassing characteristics like perfectly round, fluffed up and soft. Women my grandmothers' age did grind their own flour but nowadays everybody either buys preground flour or again takes it into a community grinder. Roti is actually quite similar to tortillas, especially flour tortillas, but distinct from naan in that both roti and tortillas are unleavened.

Well, at least the last part is somewhat encouraging, because part of the tortilla story is how daughters learn first to grind corn in the metate and develop skill in shaping and cooking tortillas. It is an important part of Mexican culture.

But further research reveals that the story for frybread is much different. Its history includes the forcing of Native Americans onto reservations and federal agents and agencies bringing (surplus, low-quality) foodstuffs (aka "commods") to the reservation, including huge quantities of (unhealthy) lard. It is testament to the creativity of Native Americans that they used the lard to develop and make (unhealthy) frybread. So frybread is a part of Native cultures, but its symbolism is ambiguous at best—younger Native children who do not know its history may simply think of it as part of their culture and good to eat. For some, frybread might reflect innovation and resilience. And for others, well, it tastes pretty darn good. For elders the memories may be tied to oppression and suppression.

And pancakes? There is presumably a story here but we don't know it and our U.S. informants don't seem to know it either.

This is all quite distressing. The tortilla is a fundamental construct in the practices and meanings associated with Mexican food and we seemed initially to find promising cross-cultural parallels (naan, injera). But then things started to fall apart.

Our research team now has its good and bad days (though we're not sure which are which). On some days we decide that the meaning of things like tortilla is tied up with our history and by relationships to both other people and other things like corn. These relationships are so interwoven that trying to analyze tortilla from a cross-cultural perspective is just not a meaningful exercise, precisely because it is so embedded in relationships.

On other days we remain tantalized by the intriguing correspondences between tortilla and naan, injera, and perhaps pizza as well (the research on pancakes has fallen out of favor). We wonder if there is some more abstract level of analyses that would be compelling, one that would avoid the messiness that our more detailed analyses revealed. We tried "bread" but it didn't seem to take us very far. "Carbs" helps us connect with rice and pasta, but this term seems so abstract and technical that we can't help thinking that it is missing the point. At least we got some nice photographs and many good meals from our cross-cultural work.

Does this story have any traction for studies of culture and thought? Suppose we are interested in science-related practices like observation, hypothesis generation, reasoning, and decision making. Are these constructs similar to natural constructs in the sense of being the same everywhere? Or are these constructs like tortillas, providing initial excitement and encouragement in our cross-cultural studies but inevitably of questionable enduring value? We are already nervous that our conception of decision making, for example, is infused with cultural particulars stemming from strong individualism and autonomy.

So here is our challenge. For some purposes it may be useful to analyze parallels between tortilla and roti, but for others a different perspective may be needed. For example, it seems at least as important to understand the cognitive and social system in which *tortilla* is situated and the system in which *roti* is situated as it is to compare tortilla and roti directly. These perspectives need not be mutually exclusive—a relational perspective may well provide the basis for comparing tortilla and roti (and for rejecting the tortilla/naan comparison). Finally, for purposes of the section that follows, we invite the reader to imagine how the cross-cultural research would have unfolded if the project had begun with pancakes or pizza or naan or frybread—it is very likely that we would have different stories.

Methodological and Conceptual Issues in the Study of Culture

It is hard not to be fascinated by culture, because it invites us to think about how things that we take for granted could be different. But the challenges of studying culture are substantial and serious. Indeed, the challenges are so serious that many cultural anthropologists do not think that anthropology can be a science.

In fact, in November 2010, the American Anthropological Association (AAA) redefined its "Statement of Purposes" from advancing "anthropology as the science that studies humankind" to advancing a "public understanding of humankind" (Section 1). This discarding of the notion of "science" as a crucial component in how the society understands itself appears to reflect a development in recent decades that has led anthropology away from science (where at least parts of it were firmly rooted) toward humanities and postmodern reflections of ethnographic descriptions as a literary genre. Although several sections within the AAA still pursue scientific goals and approaches and fiercely object to the change in the AAA mission statement, they may be losing ground.

Typically the critics of a science of anthropology are postmodernists who believe that objectivity is so elusive that researchers can only write about their experience of another culture without any pretense of describing the culture itself. This is a view that is not easily dismissed, and the burden of proof is on the researchers who believe that a science of culture is possible.

Our approach, at least for now, is going to be to try to itch where we *can* scratch. In this section we're going to focus on three general sorts of problems/issues associated with cultural research: (1) outright bias and deficit models, (2) what we call the "home-field disadvantage," and (3) our (by now) common theme that research reflects the culture of its practitioners.

Outright Bias

Early research, conducted during the colonial era by colonialists, was, in effect, a tool for colonialism. The "other" peoples, often explicitly referred to as "primitive," were seen as inferior. What better excuse (besides wanting land) for the exploitation of other people than the idea that you were bringing "civilization" to them? Any differences were seen as deficits—for example, in the domain of religion if you were not a Christian, you lacked knowledge of the one true God. You might justify almost any act in the interest of saving someone's immortal soul.

Settlers who came to the New World saw the land as pristine but in need of "improvement" in the form of clearing trees and farming. They failed to recognize and denied that the Native Americans they encountered engaged in agricultural and conservation practices (e.g., controlled burns on prairies) and that they were observing an anthropogenic scene, one rich in biodiversity. One justification for taking the land was that it wasn't "being used"; the other was that Native Americans represented an inferior race and were also in need of improvement ("save the man, kill the Indian").

These common assumptions of colonialists carried over into research observations. Let's start with a simplified example. Suppose I know how to do one hundred things, that you know how to do one hundred things, and that our skills overlap 50 percent (we have fifty skills in common). Next assume that I am in a position of power and decide to make up a test of ability by randomly selecting twenty of my one hundred skills. When I give you the test you'll likely score around 50 percent, which is clearly inferior to my 100 percent. Of course if you were in a position of power and made up the test it would be a different story. But I can't test for your unique skills because they are likely irrelevant to my life and I lack them. As long as I'm in a position of power, I define what skills are relevant and I might even use my superior performance to justify my position of power.

This example is transparently oversimplified but it is neither off the mark nor misleading. For example, ability testing in the form of intelligence tests was used in the early twentieth century to keep unwanted immigrant groups out of the United States without any consideration of whether the test items were appropriate for the group in question. Cole and Scribner (1974) describe how various tests of reasoning and other cognitive abilities developed in the West (mostly the United States) were transported elsewhere without regard to local context, with the inevitable result that children and adults in the other culture performed worse than Western children and adults.

There were a few bold anthropologists who, for example, in their studies of other cultures, marveled at the people's knowledge of plants and their medicinal uses. These observations led them to question the idea that these people were intellectually inferior, but these anthropologists were the exception rather than the rule.

Cognitive psychologists have the term "confirmation bias" for this sort of strategy of documenting deficits. Confirmation bias (Wason 1960) refers to a pattern of gathering information that will either prove one's hypothesis or perhaps be ambiguous, as opposed to seeking information that might directly undermine one's hypothesis. Here's a very simple task that cognitive psychologists have used: I have a rule and certain number sequences obey the rule (and some do not). Here are your instructions: "The number sequence 2, 4, 6 follows the rule, as does the sequence 24, 26, 28. You can give me sets of three numbers and I'll tell you whether or not they follow the rule. Let me know when you think you know the rule."

If you're like the typical participant you may think you already know the rule and might test your idea by asking whether 6, 8, 10 follows it (it does). After a few more tests you might decide that the rule is "even number incremented by 2," at which point I'll tell you that you are wrong. Other participants may test part of their hypothesis by asking about 11, 13, and 15 (yes, it follows the rule) and then conclude that the rule is incrementing by 2 (wrong again), or by asking about 42, 44, 48 (yes) and then conclude that the rule is increasing even numbers (wrong).

The rule, as you may have guessed by now, is any increasing sequence of integers (though we could be even more sneaky and allow the rule to include decimals). The point is that it is easier to formulate tests that are consistent with your hypothesis or theory than to come up with tests that undermine your theory. Ask any researcher whether they would rather do a study that would either support their own theory or be ambiguous or a study that would support a rival theory or be ambiguous. We're betting that the former option will get a lot more votes than the latter (and that this will be reflected in actual practice). In the case of cultural research, if you're already convinced that you come from a superior culture, chances are you'll do research confirming that impression.

For purposes of this chapter we are going to assume that the most egregious and most transparent misuses of cultural research have been called out (for a careful, thorough, clear, and insightful analysis see Cole and Scribner 1974). Our attention will now shift to the more subtle and perhaps more pernicious forms of cultural bias that grow out of what we refer to as "the home-field disadvantage."

The Home-Field Disadvantage

We use the term "home-field disadvantage" to refer to the disadvantage inherent in research that takes a particular cultural group—or that group's performance—as the starting point or standard for research, especially for cross-cultural research. Given that psychological research has been mainly conducted in the United States, it is almost always the case that the beginning point consists of results obtained in labs at major U.S. research universities. It is not obvious that this is a disadvantage in itself, other than limiting the presumed generality of results. Nonetheless, we argue that the home field is a serious handicap and leads to misleading cultural differences, often in the form of deficits (see Medin et al. 2010 for further details).

It may be useful to distinguish among senses of home-field status. First, researchers themselves often come from only one of the cultural groups being compared (*in-group* as home field). Second, historically these researchers often occupy a position of power and authority relative to other groups being compared (*power* as home field). The fact that white American males do a great deal of the research on culture is an example of these two paths to home-field status. Third, research *design, evaluation,* and *analysis* often originate (or originated) with only one cultural group, whether or not the researchers themselves are from that group (*starting point* as home field). The fact that the research protocol is often originally designed, evaluated, and analyzed using primarily white, U.S., university undergraduate students as participants is an example of the first (starting point) and third paths to the home field.

All three paths to home-field status asymmetrically influence the *psychological distance* between researchers and the cultural groups being studied. Psychological distance refers to how subjectively (*psychologically*) close or distant an event or person is across a range of measures of distance, including time, space, and personal identity (Trope and Liberman 2003). These three paths to home-field status push in the same direction such that researchers are psychologically closer to (1) their cultural *in-group*; (2) the cultural group they take as a *starting point* for research design, evaluation, and analysis; and (3) the dominant or majority culture group.

Home-field status—with its relationship to psychological distance—in turn contributes to at least three distinct (though overlapping) disadvantages. These are the more subtle dangers that we warned about. First, it affects whether a group's cultural practices and beliefs are seen as normal or deficient (the problem of *marked vs. unmarked culture*). Second, it affects the degree to which cultural groups seem more uniform and easy to essentialize versus diverse and multifaceted (the problem of *homogeneous vs. heterogeneous culture*). Third, by virtue of the process by which research stimuli and methods are selected, home-field status makes it likely that cultural differences will be found that have no basis in reality. For reasons that will become clear we refer to this as the problem of *regression toward the mean*. We now describe each of these sources of bias in some detail.

Marked versus Unmarked Culture

At a typical psychology convention, if research participants are mentioned at all, the speaker most often says "people," because it is safe to assume that they are talking about undergraduate students at a U.S. university (and most likely to students taking the Introduction to Psychology course), along with the additional unmentioned demographic and cultural characteristics that go along with being a student at a U.S. university. There are many reasons why this particular cultural group is unmarked, of course, including all three paths to the home field we just mentioned. One problem with this cultural group (i.e., the standard research population) being unmarked is that its peculiarities go unnoticed. Instead, the characteristics of this group are the characteristics of "people," implicitly taken to be representative of humankind, and providing insight into universal, culture-free human psychology. As we noted in chapter 1, (1) claims about human psychology and behavior are largely based on narrow samples from Western societies; and (2) the still relatively meager amount of cultural research nonetheless suggests that findings do *not* readily generalize to other samples and that undergraduates may be especially atypical of the world at large (Henrich, Heine, and Norenzayan 2010).

One does not have to leave the United States or even these same college campuses to undercover the issue of generalizability. Snibbe and Markus (2005) report differences in judgment and decision making between middle-class and working-class participants, and Lehman and Nisbett (1990) documented differences in reasoning among undergraduate social science majors and humanities and natural science majors—differences that developed over years in college, suggesting that majors participate in different (academic) cultures.

The problem of markedness may persist among cross-cultural researchers, though in less direct ways. Although we may carefully mark the demographic characteristics of our research participants, the cultural specificity of the normally unmarked group's *practices and beliefs* are not so easily marked.

Consider the following example from Medin's research with Scott Atran (Atran and Medin 2008). One of the things these investigators looked at in their cross-cultural studies concerns the so-called *diversity principle* in reasoning about categories. To see how this principle works, readers can participate in an experiment. Suppose you know that Disease A affects river birch and paper birch, and Disease B affects white pines and weeping willows. Which disease do you think is more likely to affect all trees? If you give problems like these to University of Michigan undergraduates, over 90 percent answer Disease B, the disease that affects white pines and weeping willows (López et al. 1997). If you ask them why, they say something like this: "Well, Disease A could be just a birch thing, and if it happens for trees that are this different—as different as white pines and weeping willows—it's more likely to affect all trees." This is the diversity principle in categorization and, at least initially, Atran and Medin thought it might be universal. But when they tested it with Itza' Maya agro-foresters in Guatemala they found *below*-chance diversity responding (López et al. 1997).

The problem comes from starting with what you know. The "natural" question for Atran and Medin to ask was, "Why do the Itza' *fail* to show diversity?" because diversity seemed like *the* only reasonable strategy. As it happened, Atran and Medin were also simultaneously doing studies involving another population consisting of US adults who knew a lot about trees—people such as parks workers and landscapers (Proffitt, Coley, and Medin 2000). Like the Itza', many of these experts also showed below-chance diversity responding. For example, when presented with the disease scenario previously outlined, thirteen out of fourteen parks workers picked Disease A—the disease that affects river birch and paper birch—as more likely to affect all trees. Their typical explanations were causal and ecological. They said such things as: "Well, first of all, birches are incredibly susceptible to disease. If one of them gets it, they'll all get it. Secondly, they're very widely planted as an ornamental, and they're widely dispersed naturally; so there would be plenty of opportunities for that disease to spread."

Note that had Atran and Medin begun their studies with Itza' Maya or tree experts, the deficit thinking would have been reversed: "Why do college undergraduates fail to show causal and ecological inductive reasoning?" Even the notion of "expert" depends on one's perspective. The tree "experts" could identify roughly 90 percent of the trees in the Chicago area, but where Atran and Medin work in Guatemala, nearly everyone can identify nearly all the trees. Itza' Maya might find it very strange for us to call our tree "experts" experts. Whether the landscapers and park personnel are experts or Northwestern undergraduates have a nature deficit disorder depends on where you start.

Homogeneous versus Heterogeneous Culture

Social psychologists have established that relative to in-groups, people (people = undergraduates) tend to see out-groups as homogeneous. Here's a concrete example. A few years ago Louis Gomez and the first author taught a class where one of the readings concerned negotiations over an Arab-Israeli conflict in the Middle East. At one point several of the students suggested that what we really need to know is the set of beliefs that all Muslims share. We then asked the students what that list would look like for Christians, only to be told that Christians are simply too diverse in their beliefs for one to make meaningful generalizations. To the students' credit, they quickly appreciated the inconsistency of their intuitions and backed away from wanting to know "what Muslims think."

If out-groups are seen as homogeneous, then it will seem natural to aggregate over broader groups. Our earlier example of Chandler and Lalonde's (1998) research looking at suicide rates among First Nations people in British Columbia illustrates how misleading results can be if the level of aggregation is too broad.

Regression toward the Mean

Another home-field disadvantage in cross-cultural research comes from knowing the "sweet spots." Suppose we begin with some psychological result or effect that has already been established using a North American sample, use experimental materials that other people have been effectively using (with that same sample), or follow methodological procedures that are already well honed within our own cultural group. If we then translate and transport these materials and methods for a cross-cultural comparison, chances are we are likely to find that the effect either is weaker or disappears. This may well happen even when researchers work to translate the materials into local languages and otherwise take steps to ensure that the assessment tools in use are locally meaningful and familiar. A common interpretation of this pattern of results is that we have discovered a bona fide cross-cultural difference, but such an interpretation may be problematic because of regression toward the mean.

Perhaps the easiest way to understand regression to the mean and its implications for cultural comparisons is with a concrete example. Suppose that we decided to study "sense of humor" and developed a set of jokes in North America and then (after proper translation) tested them in another culture. We could be quite sure that people in the other culture would not find them as funny as people in the home culture. But no reasonable person would want to conclude that people in the other culture had a worse sense of humor. We recognize that jokes have been *selected* for funniness and that funniness is determined by knowledge, values, and other sorts of individual and cultural variables. But what holds for jokes also applies to any measurement tool when it is selected and developed in one culture and applied to another.

More broadly, research findings are not simply measures of psychological phenomena—they are also the result of an interaction between psychological phenomena and the procedures and materials tailored to bring them out. Through trial and error, or active intention, research materials are selected according to how well they fit the participant population in order to produce a particular effect. For example, the most widely cited example used to show framing effects in decision making—Tversky and Kahneman's Asian disease scenario (1981)—also happens to produce larger effects than almost all other framing scenarios that have the same abstract structure (Kühberger 1998).

Looking for items that bring out some effect of interest makes perfect sense and represents skillful research, but it has inimical consequences when used asymmetrically in cross-cultural comparisons. Because the particular stimulus materials have been selected to bring out particular effects with a particular cultural group, *regression toward the mean* predicts movement away from this exceptional performance toward the less exceptional, even if the second cultural group would display the same psychological phenomenon equally well (or better) using a different set of stimuli that had been tailored for success with them.

Of course, this does not always contribute to a perception that the second cultural group is deficient. The Asian disease scenario, for example, is designed to demonstrate deficiencies (i.e., deviation from normative models of decision making). In such cases, regression toward the mean suggests that other cultural groups will appear *less* deficient rather than more deficient. In general, the fact that research methods have been selected to "work" with a particular cultural group (usually American, usually university students) gives that group a privileged status that is not shared with other cultural groups, whether this privilege makes the original group look better or worse or just different.[1]

Is there a cure for the home-field disadvantage? Well, there are some steps that will help. For example, one could recognize that research findings represent a relationship between a particular cultural group and the study design, and that, therefore, when these same designs are applied to a different population, some kind of flattening of performance (regression toward the mean) can be expected. Here's one additional example. In developmental studies some clever researchers have found that it is easier for young children to answer a question if you bring in puppets who "don't know the answer," rather than answering to the adult experimenter who presumably does know the answer (Herrmann, Waxman, and Medin 2010). But in our studies with Native American children who may have much less experience with puppets, the use of puppets may have been more of a distraction than a help.

Another way to address the home-field disadvantage is to include researchers who are members of the "away" culture. We'll have more to say about this later, but for now we remind you of Elissa Newport's comment about research on deafness including a "token" deaf undergraduate research assistant. For this strategy to work there must be power sharing.

A third approach, which is particularly challenging, is to do one's best to study the phenomenon of interest on the terms of the culture, or cultures, being studied. Maurice Bloch, a seasoned, wise anthropologist, describes this as studying a culture from an internal versus an external perspective. For example if one were studying cultural differences in emotions, it would be a mistake to start with English emotion terms and try to identify their counterparts in another culture, since this presumes part of what one wishes to study (Boster 2005). In other words, there is much to be gained by changing the starting point of investigation and the home-field disadvantages that come with it.

The Home-Field Disadvantage: Summary and Implications
We end this section with a thought problem: what would the field of psychology look like if its beginnings had been in China? Psychology may be the least international of all the social sciences, and it certainly is more

parochial than any of the natural sciences. Therefore, there is plenty of reason to think that a different point of origin would have led to a different current state of the art (see also Medin and Bang 2008).

Once one starts along this line of thought, it is easy to generate variations. To what extent do our current understandings of learning, cognition, and decision making depend on the gender, ethnicity, socioeconomic status, political leanings, and cultural identification of the scientists who are conducting psychological research? For example, what if the field of psychology had been considered as the sole domain of women? Would our theories and empirical generalizations look much different? There is reason to think that they would (see Eagly 2005 for an example of research on leadership).

Envisioning the various ways the field might have evolved under various forms of restrictions is sobering, because it underlines the problems of insights lost by virtue of a lack of diversity. Note that the costs of limited perspectives not only fall on the field of psychology per se, but also may be devastating for the populations affected by the research (e.g., immigrants, women, people of color, etc.), especially when biased assessments hold the authority of science.

Trying to take a different perspective and succeeding in doing so may be two different things. Temple Grandin, who argues that her autism helps her take the perspective of animals, provides a telling example in her book with Catherine Johnson (2005). The story comes from ethologist Ron Kilgour. A person had a pet lion that he was shipping on an airplane. Someone suggested that providing a pillow would ease the stress of the trip. They gave the lion a pillow, which turned out to be a disastrous decision because the lion ate the pillow and then died. Temple Grandin read the story and said to herself, "Well, no, he doesn't want a pillow, he wants something soft to lie on, like leaves and grass." The trick is to think like a lion, not a person.

In some cases indigenous people have evolved a system that facilitates perspective taking. For example, the Itza' Maya (of Guatemala) belief system promotes perspective taking with respect to the rain forest. They see the forest as inhabited by spirits (the Arux): not abstract entities whose presence is inferred, but spirits who sometimes are visible and are known to drink from the water vine (*bejuco de agua*) and enjoy the flavor of the food they obtain from the allspice tree (*pimienta gorda*). When we asked Itza' elders to rank-order twenty-one species of plants from the point of view of the Arux, the best predictor of their rankings was ecological centrality (Atran and Medin 2008). In short, they see the Arux as guardians of the forest. These rankings are correlated with but different from the rankings that they provide for themselves (their own needs enter in) or from their view of what God's perspective might be.

Other groups in the area also believe in the forest spirits but don't see them as relevant to the forest per se. Their agro-forestry practices are much more destructive than those of the Itza' (see Atran et al. 2002). It appears, then, that the presence of the Arux helps the Itza' take the forest's point of view.

So far we've suggested that one should do one's best to take multiple perspectives in cultural research and that, at the same time, it may be difficult to do so. One solution is to seek research partnerships from within the cultural groups being studied. In our studies of agro-forestry in Guatemala there were no obvious academic counterparts, but there were plenty of indigenous biological experts. We never conducted an interview or ran a task without first consulting with local Maya elders. They were interested in what we were doing and we benefited immensely from their efforts to help us understand the data we collected.

We have been able to establish research partnerships with Native American institutions for our work conducted on the Menominee reservation in Wisconsin and in Chicago over the past decade. For example, our grants have involved Northwestern University, the American Indian Center of Chicago, a tribal college on the Menominee reservation, and (more recently) the University of Washington and the Language and Culture Commission of the Menominee Nation. It is to the credit of the National Science Foundation that these partnerships do not involve subcontracts from Northwestern University to tribal institutions but rather parallel budgets with a principal investigator at each site.

It is common to say that one has to start somewhere, implying that a single locus is a logical necessity. One person can't be two places at once, but a research team can. Further, by virtue of research partnerships and self-consciously developing research tools and methods simultaneously in multiple contexts, one can go a long way toward limiting the asymmetries that seem endemic to cultural research.

One byproduct of research partnerships is that they reduce the asymmetries in cultural research and they provide multiple perspectives. Having our research approved by Northwestern's Institutional Review Board is only one of several steps. AIC community members and the Menominee Language and Culture Commission must also approve our research, and they have an opportunity to bring their values to bear on the project.

The design of our studies also is based on an understanding of appropriate research methods for working with American Indian communities. There is a long history of research in Indian communities that has often not been in their best interest, a legacy that has made many Native communities rightly suspicious of research. Over the years indigenous researchers themselves have worked to develop appropriate research methods and criteria (e.g., Guyette 1983; Hermes 1999; Kovach 2010; Mihesuah 1998; Smith 2006).

One example of benefits from this research partnership comes from the first author. When he began doing work with Menominee participants, it seemed perfectly fine for students to complete their dissertations by paying participants generously ($20 per hour), seeking appropriate tribal approvals, and employing Menominee research participants whenever possible. Now it doesn't seem so fine. Spending a total of a few thousand dollars on participants seems a pretty modest cost relative to the value of a PhD. It's also clear that tribes or intertribal communities such as the American Indian Center have a great need for indigenous scholars who can apply for grants, build research infrastructure, and bring relevant expertise to bear on policy issues. One of the most promising aspects of our research partnership is that it does appear to be facilitating Native scholars' seeking advanced degrees.

The second author has benefited from these partnerships in a different sense. Among many other things, we note that she has gained access to the norms and expectations of powerful research institutions like Northwestern University. This has proven useful to her resilient navigation of the academic system and afforded her the opportunity to support other Native students' navigation of these systems. Further, these research partnerships have created new policies and procedures in doctoral work. For example, the second author's dissertation committee included two tribal community members, thereby raising the graduate students' academic accountability to the communities and recognizing the role of community members in shaping the doctoral work of the graduate students. This inclusion of community members on her committee also recognized that expertise within particular communities is not necessarily reflected by educational attainment status.

Cultural Comparisons Are Challenging

One of the authors is fond of saying that two things can happen when one does cultural comparisons and neither one is good news. First, one can invest the time and trouble, addressing all the pitfalls we have described, and compare two cultures but find no differences. In that case our costly and time-consuming efforts would have served only to verify what our colleagues already (thought they) knew—that the results they had obtained with U.S. samples would generalize broadly. We *might* be able to find a scientific journal that would publish our finding of lack of differences.

The other possibility is that we invest the same time and trouble and *do* find reliable cultural differences. In a sense this is even worse news, because now we are challenged to figure out *why* we found differences (and it better not be because of any of the pitfalls we've been discussing). The logic of cultural comparisons is just the opposite of the logic of a controlled experiment. In a well-designed study, there typically is an experimental condition and a control condition that differs from it only with respect to a single factor of interest. Then when you find a reliable difference it seems obvious that the single factor is what's responsible for it. But cultural comparisons inevitably confound an enormous number of factors. For example, when U.S. undergraduates showed diversity effects and Itza' Maya elders did not, the responsible factor could have been age, being monolingual versus bilingual, income level, family size, the particular plants and animals one typically sees, type of government, literacy rate, language, familiarity with hypothetical questions, knowing the Michigan fight song, knowing how to ride a bicycle, previous travel experience—and the list could go on and on. There is a sense in which comparing two cultures divides the world in two and any of the ways the two halves differ is potentially relevant.

We may be guilty of painting a somewhat gloomy picture. The nature of the study may transparently rule out many factors as irrelevant. Knowing how to ride a bicycle presumably has nothing to do with whether or not you show a diversity strategy in inductive reasoning. Well, come to think of it, maybe we shouldn't be so sure. There is work showing that having a bicycle (a tool that may allow someone to attend an otherwise inaccessible school) is correlated with the loss of biological knowledge (Zent 1999), and not using diversity on an inductive-reasoning task appears to be related to having biological expertise. So you see the problem.

There are at least four strategies for dealing with this problem. One is to bring in a third group that is similar to one group in many ways but also similar to the other group in some respects. This is what we did (by accident) in finding that U.S. biological experts reasoned in the same way as Itza' Maya elders, thereby ruling out a host of factors. This sort of triangulation strategy can be effective if you are lucky, and adding more comparison groups can also help if (again, with luck) they form a coherent pattern.

A second strategy is just to ignore the problem and make your best guess as to what is responsible for the difference that you observe. Studies on language and thought sometimes adopt this strategy by assuming that language differences are responsible for the observed differences. This isn't as rash as it sounds, because the measures have been selected on theoretical grounds linking the measures to language differences.[2] This strategy is also in use in a significant amount of recent cultural research comparing the cognitive consequences of Western individualism and Eastern collectivism. Even if individualism versus collectivism turns out not to be the critical factor, there is, at a minimum, an accumulating body of evidence showing that ways of thought are not universal.

A third strategy is to note that not every variable one could name has the same status as explanatory variables. To give a simple-minded example, one might note that participants in one sample have bigger homes, drive more expensive cars, are more likely to own a vacation home, more likely to make foreign trips, more likely to have a gardener, and so on, than another group. But instead of dignifying each of these variables, one would presumably consolidate them under the superordinate term "high income." It may not have been obvious that having a bicycle is linked to having less biological knowledge in a culture under examination, but once biological expertise is identified there could be any number of measures of it and its correlates.

The fourth strategy we'll suggest is somewhat similar to the third and perhaps the second as well. We've called it a systems-level approach, and the idea is to conceptualize a culture as a complex system of related variables rather than independent variables. On this view the kinds of measures that one collects in a typical study may tend to point to some common themes or abstract ideas that are important to a culture. For example, if politeness and respect are important for the functioning of a culture, this may be reflected in rules about bowing, honorifics in the language, themes of stories for children, the practice of taxi drivers wearing white gloves, and the wording and appearance of signs in public. If one contrasts two cultures that differ in the importance of politeness and respect, one might well observe cultural differences on a wide range of measures related to this theme.[3] In short, this strategy consists of working to identify broad themes or principles that are important to a given culture, and then and only then attempting to understand culture differences. One such broad theme consists of how human beings in various cultures see themselves in relation to the rest of nature. We take up this issue in the next chapter.

7 Psychological Distance and Conceptions of Nature

It seemed like a good idea at the time.
—Ubiquitous

In 1984 Medin was visiting Emory University in Atlanta, Georgia, when the following events took place. It was a warm day and Medin had just gone for a run on the outdoor Emory track and was starting to head home to shower and relax. He turned on the car radio and almost immediately heard an emergency weather warning—a tornado had been sighted in the Atlanta area and citizens were urged to seek shelter. At this point Medin was closer to Emory University than his apartment so he turned around and headed back. Next he began to think about what he would do at Emory and thought about how cold the air conditioning kept his office as well as how sweaty he was. The cool air that would normally be welcome didn't seem like fun for someone in shorts and a T-shirt, so he once again turned around and headed for the apartment, fully aware of how unwise it was to be circling around rather than getting out of (a potential) harm's way.

Being a psychologist and seeking to avoid personal embarrassment for his foolishness, Medin began to think about how this episode might reflect some psychological principles of planning and decision making. Clearly we often make decisions without planning out all the details (perhaps because we might well end up lost in thought and never make decisions), sometimes to our regret. Two social psychologists, Nira Liberman and Yaacov Trope at New York University, must have had similar intuitions, but they were creative enough to know what to do with them. They published a seminal theoretical paper (Trope and Liberman 2003) laying out what they called "construal-level theory" and reviewing a range of clever studies showing its power.

Here's the gist of their theory (with apologies for skipping over nuances). First, we need to call on the straightforward idea that aspects or features of a situation that people might think about can differ in how abstract or concrete they are. For example, "try to catch a fish" is more abstract than "use a spinning reel to cast a red and white daredevil lure in shallow water to fish for smallmouth bass"; or "aim to win an Olympic gold medal in the marathon" is more abstract than "run 26 miles 385 yards next Saturday at Allerton Park," which in turn is more abstract than "put on your sunglasses and New Balance running shoes, stretch for five minutes, and then step outside and head north on Sheridan Road." A correlated distinction is that some features focus on the whys of actions and others are concerned with the hows of action. The hows are generally more concrete than the whys.

The next key assumption is that in psychologically close situations people attend relatively more to concrete features than they do in psychologically distant situations. This notion makes perfect sense—if you're going out for dinner tomorrow night you'll likely be thinking about where you'll go, what kind of food you want, what time you'll go, and perhaps about whether or not you need a reservation. But if you're going to be on vacation a year from tomorrow night, you're probably not thinking about any of these things, even though you may be equally certain that you will eat out then. This all seems straightforward, but Trope and Liberman have shown just how powerful these ideas can be.

The boldness of their theory can be seen in their willingness to test both the limits and the implications of psychological distance. For example, Liberman and Trope have examined not only temporal distance but also spatial distance and social distance. In fact, they have even mapped their framework onto probabilities by assuming that events with low probabilities of occurring are treated as more psychologically distant than events with high probabilities of occurring.

Please excuse the following slight diversion as we tell you about one of the studies reported by Rottenstreich and Hsee (2001). They had forty Rice University undergraduates fill out a questionnaire that included a probe asking them to imagine that they could choose between receiving fifty dollars in cash or the opportunity to "meet and kiss your favorite movie star." Twenty of the students were told to imagine that there would be a lottery and that they would have a 1 percent chance of winning and being able to choose. The other twenty were asked which option they would choose with no uncertainty introduced. Rottenstreich and Hsee found that 65 percent of the students preferred meeting and kissing the movie star when the odds of winning were 1 percent, but only 30 percent preferred that option when it was certain!

Rottenstreich and Hsee (2001) have their own interesting interpretation of this result, but we mention it to illustrate construal-level theory. According to the theory, when the choice is part of a long-shot lottery, it will appear to be psychologically distant and higher-level, and abstract features will get more weight. The idea of meeting and kissing your favorite movie star may be more appealing at a distance than it is close up. In the lottery case, you may not even have thought of who your favorite movie star is, and when you do you may realize that they are not someone you would want to kiss. Or you might start to think about more concrete issues, like how your significant other is going to feel about you kissing a movie star. So the distant desirability may fade as you start to consider the concrete realities.

We're probably all familiar with situations where we find that the desirability of some action changes as we get closer to the event. For example, you might agree to attend some event that will take place a month in the future because it sounds like fun, only to find yourself with much diminished enthusiasm closer to the event when you have to think about finding a baby sitter, the long drive to the event, and a bunch of other details you hadn't considered before.[1] One way of describing this pattern of thought is to say that at a distance, we focus on desirability, but as we get (psychologically) closer we shift our focus to feasibility (Liberman et al. 2007; Trope and Liberman 2000).

Construal-level theory leads to a wide range of other predictions that have been supported (see Liberman and Trope 2008 for a review). For now we are going to focus on two related effects having to do with distance and judgment. A great deal of research in social psychology has examined the attributions people make about the actions of others. One contrast that has received a lot of attention is what social psychologists call "situational versus dispositional attributions." When you ace a test, is it because you are smart or because the test was too easy? When a lifeguard rescues someone caught in an undertow, is she heroic or "just doing her job"? Or imagine that an elderly man gets on a bus and a seated person gets up and offers her seat to the man. Does that indicate to you that the woman is polite and respectful or is it just what anyone would do in this situation and this woman just happened to be the first one to notice the elderly man? These are judgments that people make all the time. So-called "dispositional attributions" are when you explain behavior in terms of qualities of the actor, and "situational attributions" are when you explain the behavior as determined more by the situation.

You can test your understanding of construal-level theory by asking yourself how psychological distance affects attributions. And you are correct if you decide that the theory predicts that psychological closeness promotes attention to the context and situational attributions; in contrast, psychological distance should be associated with more abstract dispositional judgments. This prediction is well supported (e.g., Henderson et al. 2006).

We also want to draw your attention to the observation that being in a position of power is also associated with psychological distance. A position of power appears to be associated with a reduced ability to take into account another person's perspective (Galinsky et al. 2006). One might also predict that people in a position of power see others as psychologically more distant and are more likely to make dispositional attributions. Our social psychologist friends tell us that this particular prediction hasn't been tested (yet).

Implications and Applications

This is all well and good, but what does construal-level theory have to do with the nature of science, science education, or culture? Well, first of all, there do appear to be chronic cultural differences in perspective taking and sensitivity to context, with Westerners being more dispositional and Easterners more situational (Nisbett 2003; Wu and Keysar 2007).[2] The potential relevance to science and science education comes when we consider the psychological distance of human beings from the rest of nature. Specifically, we are going to argue that there are substantial cultural differences in framework theories or epistemological orientations toward the natural world and that these differences affect science-related practices. The link to construal-level theory is that one key difference among framework theories is along the dimension of psychological distance. One implication or question worth pondering is whether the notion that scientists must be "detached" and not get too close to their subject matter is just a cultural preference for psychological distance. We will see that this distancing has important cognitive consequences.

In much of the following we're going to focus on Native American and European American children and adults, and we're going to analyze psychological distance as one component of differing cultural models of nature. There are stereotypes about Native Americans being "closer to the land" than European Americans. (We apologize for using such broad, distancing categories and the invitation to make dispositional attributions about them. We have also done some work with an Amish community in Wisconsin and we'd bet a lot that measures of distance would also find Amish farmers are "close to the land.") We will soon present evidence that is consistent with this view.

Just one word of warning. In terms of our own cultural practices for doing science, we're starting out with a fairly typical Western science orientation by identifying a single key variable and tracing its correlates. Later on we'll shift to a more relational perspective having more similarity to Native science than Western science.

Cultural Models of Nature

The elder member of our research team likes to say that cognitive psychologists go through developmental stages in their research. In Stage 1 they make a number of simplifying assumptions about what they are going to study and how they are going to study it. In the second stage they produce some (we hope) interesting findings. But in Stage 3 they see these results as of limited value, precisely because simplifying assumptions have been made. They then argue that things are more complicated. It's a little like eating a cake and then denying that there ever was a cake.

We're going to do the same thing here. We're going to assume that cultural differences on conceptions of nature are largely captured as differences in psychological distance from (the rest of) nature. We report a variety of measures of this hypothesis, all of which are nicely convergent. In chapter 8 we describe a series of studies on the cognitive consequences of differences in psychological distance that also supports this general framework. But then we're going to own up to what we see as limitations of psychological distance and introduce some additional notions about cultural orientations toward nature. We're not going to deny that there ever was a cake, but we will suggest that, by itself, cake is not a complete meal.

Measures of Conceptions of Nature

There are many, many facets to culture, and we should also expect that there are many, many facets to cultural conceptions of nature. These conceptions may be manifest in explicit statements that are readily elicited from informants, observable in public cultural artifacts, or either implicit or explicit in practices.

Earlier we mentioned the Itza' Maya in Guatemala, their belief in the Arux, and the belief of Maya elders that the Arux are guardians of the forest. They also refer to the forest as the Maya house, and some elders say explicitly that when the forest is gone the Itza' Maya will no longer exist as a people (Atran et al. 2002). Early on in our research with the Itza' we asked Itza' elders how they learned about the forest. Almost to a person they said, "I walk in the forest."

We also commonly tape-recorded our interviews, which were often conducted in people's homes. It was always striking when we played back our recordings how many animal sounds (mostly domestic animals) were present, sounds that we had tuned out to listen to our informants. For the Itza' Maya the rest of nature is psychologically close.

We employed a wide range of measures in these Guatemalan studies (Atran and Medin 2008). These included asking Maya adults to use name cards to sort animals into groups that made sense to them (and to explain their sortings), to reason about plant-plant, plant-animal, and animal-animal (ecological) relations, to rank-order the importance of different species of plants from different perspectives (including that of the Arux), and to predict how other cultural groups would rank order biological kinds. We also asked Itza' living in the same area whether they helped, hurt, or had no effect on different biological kinds; collected soil samples from their agro-forestry plots; and counted the types and sizes of trees on their various plots. We'll say more about our results later—for now we just want to emphasize that employing a wide range of measures is useful if not necessary for telling a coherent story.

Contrast Itza' Maya elders with the typical Northwestern University undergraduate student. The majority of undergraduates know very little about the tall woody plants that are around campus. For example, on a survey we conducted, less than half of them indicated that they had even heard of hackberry, hawthorn, or buckeye trees (Atran and Medin, 2008). Ignorance about the last of these is especially surprising given that the Ohio

State Buckeyes are part of the Big Ten conference, and thus a frequent opponent of Northwestern athletic teams. Although we haven't collected all the relevant converging measures, our best guess is that, on average, nature is not psychologically close for Northwestern University undergraduates.

Nature and Psychological Distance

The primary focus of our research on psychological distance has been two cultural groups living in close proximity in north-central Wisconsin: rural Native Americans from the Menominee Nation and rural European Americans living in an adjacent county. Where we have relevant data we also include a third population, urban Native Americans. First we'll briefly describe our study samples.

Menominee The Menominee are the oldest continuous residents of Wisconsin. Historically, their lands covered much of Wisconsin but were reduced, treaty by treaty, until the present 95,000 hectares in total was reached in 1856. There are 4,000 to 5,000 Menominee living on tribal lands in three small communities. There are also about the same number of Menominee living off the reservation. We have made no effort to ensure that our samples are random, because our goal is not to describe the average Menominee person or estimate what or how Menominee people in general think. Instead our goal is to uncover and describe distinct cultural models of nature. A separate question is whether Menominee who live off the reservation differ from those living on it.

Despite economic incentives to the contrary, the Menominee have preserved diversity and habitat types of their forest (Beck 2005; Hosmer 1999), which is managed by a tribal corporation. They enjoy an international reputation for their sustainable forestry practices (Hall and Pecore 1995).

European American Just south of the reservation is Shawano County (population about 40,000), the other focal area for our research. The major sources of income in the town of Shawano are light manufacturing, small-scale farming, and tourist recreation, mainly in the form of hunting, fishing, boating, jet skiing, and snowmobiling. Shawano Lake is a major attraction and there are also several rivers and smaller lakes in the county. Outdoor recreation is important to many of the county residents, and many adults have fished since they were young children.

Urban Native American Our urban Indian sample is drawn from the greater Chicago area. According to the 2010 census there are about 42,000 Native Americans in this area, and we primarily recruited our sample from people who either participated in activities associated with the American Indian Center (AIC) of Chicago or were socially connected with participants. The AIC is the oldest urban Indian center in the United States and provides educational and health services and activities to a large number of Native Americans and non-Indians. Membership is open to all and the AIC is intertribal in character. Some of our participants were born and raised in Chicago, whereas others came to Chicago due to U.S. government efforts during the relocation era. Many urban Indians have relatives living on a reservation and may spend time there.

Study 1: Outdoor Practices

The first study we report (Bang et al. 2007) is based on our two rural samples. The Menominee sample consisted of sixteen adults and eleven children, and the European American sample consisted of fifteen adults and seven children. Participants were asked to fill out a survey to identify the kinds of outdoor activities they engaged in and how often they did them. The list of practices was developed from focus groups of members from each community. There were a total of thirty-three different practice types on the survey form, and we left space for additional practices to be added.

The practices were wide ranging and included hiking, berry picking, hanging up laundry, barbequing, boating, collecting firewood, yard work, camping, dirt biking, playing baseball, hunting, fishing, and attending an outdoor powwow. Our interest was in the relative frequency of practices in which nature was foregrounded versus backgrounded. For example, we thought of activities such as forest walks and fishing as having nature foregrounded and activities like hanging up laundry and playing baseball as having nature backgrounded. Of course we asked research assistants who knew neither our intuitions nor our hypothesis to make this determination.

Coding

The thirty-three practice types were collapsed into three main categories indicating the place and role of nature, or some aspect of nature, within the practice. To ensure inter-rater reliability we had four coders (two Native American and two European American) assign each practice to one of three categories (foregrounding nature, backgrounding nature [e.g., baseball], or intermediate). We then mapped these categories onto distances (background = 1, intermediate = 2, and foreground = 3) and calculated the inter-coder correlations. These correlations ranged from +0.73 to +0.86. This demonstrates that our four coders generally agreed with each other. We were prepared for the possibility that our Native American coders and European American coders might differ in their judgments. In that event we were going to add more Native and non-Native coders and make that a variable for analysis. However, there was no suggestion of group differences in coders.

Given the good pattern of agreement, we next averaged the ratings across coders. Practices with mean scores of equal to or less than 1.75 were assigned to the category of nature backgrounded; practices with mean scores greater than 1.75 but less than 2.75 were considered intermediate; and mean scores equal to or more than 2.75 were assigned to the foregrounded category. Consensually, backgrounding practices included laundry, playing sports, dirt biking, snowmobiling, boating, cooking, and work. Foregrounded practices included forest walks, berry picking, fishing, hunting, trapping, gardening, collecting ginseng, wild ricing, maple sugaring, harvesting milkweed,[3] collecting plants for medicine, sitting outside, fire circles, camping, and using sweat rocks.

Results

The results are summarized in figure 7.1. European American participants were more likely to report practices in which nature serves as background, and Menominee participants were more likely to report practices in which nature is foregrounded. The interaction of culture and distance was highly reliable. Children's reported outdoor practices showed the same patterning of cultural differences. Children reported engaging in outdoor practices more than adults, but there was no interaction of age and culture. Overall, compared with rural European American children and adults, Menominee children and adults spend more of their time engaged in outdoor practices in which the natural world is foregrounded and less time engaged in practices where nature is backgrounded. By this measure nature is more psychologically close for Menominee participants.

We also have collected outdoor practice data with our urban Indian population. For both children and adults, urban Indian practices parallel those of rural Menominee rather than rural European Americans.

Before we go on to the next study there's one thing we need to clarify. As we have said or implied (or both) before, we are *not* suggesting that one group is closer to nature or that their practices bring them more into contact with nature. To the contrary, everyone is a part of and immersed in nature all the time. Instead, our data concern subjective or psychological distances. When we asked our coders to decide whether a given practice foregrounded or backgrounded nature, we were inviting them to adopt the (Western) stereotypic definition of nature that includes the absence of human artifacts and domestic animals. For this reason it might have been the case that our Native American coders wondered about their assignment, but either they share this stereotype or they were willing to play along with us.

Study 2: Distance in Discourse: Values and Learning Goals

You could call this an "accidental" study, at least with respect to psychological distance. Our original focus was on values and orientations toward nature. To study these orientations we borrowed from Kellert's (1993) typology of values mentioned in chapter 5 (e.g., utilitarian, ecologistic-scientific, aesthetic, moralistic, dominionistic).

The relevant data come from interviews with rural Menominee and European American adults. Each interview lasted approximately an hour and was structured around a series of questions about (1) the interviewee's

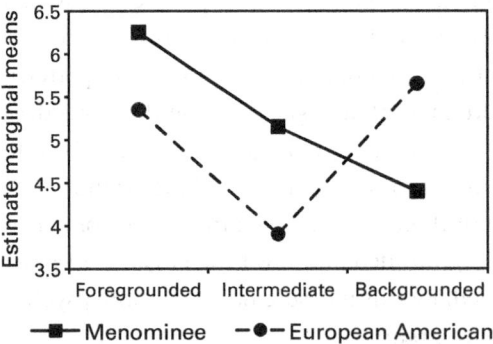

Figure 7.1
Means of reported practices by category and community.

characteristic outdoor activities and particular experiences (Study 1 just reported), (2) ways in which children should and do learn about nature, and (3) goals and methods appropriate for science instruction. Our focus was on two questions that came at the middle and toward the end of the interview. The first was "What are the five most important things for your children (or your grandchildren) to learn about the biological world?" The second question, asked in the context of discussing science education, was "What are four things that you would like your children (grandchildren) to learn about nature?" There were twelve adults each from the Menominee (seven female, five male) and European American (six female, six male) communities involved in this study. The coding scheme (to be described next) was applied to these two questions.

Coding

We expanded on Kellert's original typology in several ways. A dimension was added corresponding to subjective proximity to nature. For example, we created a distinction between "distant utilitarian" (e.g., "the wood is used in construction") and "personal utilitarian" (e.g., "we eat deer meat"). We also added categories corresponding to more abstract knowledge, social value, respect, and liking (e.g., "I want my children to enjoy nature"), and categories that were either concrete (survival skills, as expressed in "I want my child to be able to recognize poison ivy") or referred to relational stances toward nature ("everything is related") or to intergenerational values and goals (see Holistic, Spiritual, and Traditional in table 7.1).

Although we provide our full set of coding categories in table 7.1, we only report results for categories where at least a third of the participants in one or more groups gave a response consistent with the code. We also found that it was much easier to establish reliability on whether or not an orientation was present or absent than to determine whether a statement should be seen as a single instance with elaboration versus multiple instances. Our analyses are based only on the former.

Results

The dependent variable was the proportion of adults from each community whose responses were consistent with a given code on at least one occasion (see table 7.2). We focus only on the differences that are statistically reliable and those most relevant to psychological distance.

Table 7.1
Modified Coding Scheme Based on Kellert (1993)

Orientation	Definition
Holistic	A belief that everything in nature is interconnected; that there is harmony and balance in nature.
Spiritual	A tendency to access or find spiritual meanings, lessons, importance in nature. Referring to Mother Earth, Creator.
Traditional	Referring to an activity in nature as something that has occurred throughout history and should continue for that reason.
Survival skills	A belief that nature is something that can harm us if underestimated.
Moralistic	Strong feelings of moral and ethical responsibility.
Ecologistic	Precise study and systematic inquiry of the natural world and belief that nature can be understood from empirical study from a systems perspective. There is a tendency to relate species to other aspects of nature.
Abstract respect	Respect for nature.
Abstract liking	Expressing a positive but abstract attitude toward nature (e.g., "I enjoy nature" vs. the more concrete "I enjoy walks in the woods").
Abstract knowledge	Expressing learning goals abstractly (e.g., "I want to learn about nature" vs. the more concrete "I want to understand how beavers build dams").
Personal utility	The physical benefits derived from nature as a fundamental basis for human sustenance, protection, and security. The benefits are intended for the self or those in the immediate family or social network.
Distant utility	The physical benefits derived from nature as a fundamental basis for human sustenance, protection, and security. The benefits are intended for those distant from the immediate family or social network.

Almost all parents (and elders) expressed moral values and beliefs about the need to respect nature. They differed, however, in the perspective applied to these values. European American adults mentioned the need to protect nature, implying a caretaker relation with the natural world. Menominee adults, in contrast, were reliably more likely to talk in terms of "Mother Earth," and to say that they want their children to understand that they are a part of nature. This distinction between being *apart from* nature versus *a part of* nature reflects qualitatively different models of the biological world and the position of human beings with respect to it. Additional evidence of distancing comes from the finding that European American adults were

Table 7.2

Proportion of Participants Coded as Reflecting Various Value Orientations at Least Once

Orientation	Rural European American	Menominee
Holistic	0.08	0.33
Spiritual	0.08	0.42
Traditional	0.08	0.50
Survival skills	0.75	0.75
Moralistic	0.92	0.83

Note: The table only includes codes reflected in the responses of at least a third of one of the groups.

more likely to mention abstract liking and distant utility and less likely to reliably mention personal utility than Menominee adults.

There is converging developmental evidence on at least one aspect of these cultural differences. Unsworth et al. (2012) conducted structured interviews with five- to-seven-year-old rural Menominee and European American children in which each child viewed several pairs of pictures of plants and nonhuman animals and was asked how or why the species (e.g., raspberries and strawberries) might go together. Only Menominee children spontaneously mentioned personal utility. Other cultural differences relevant to consequences of psychological distance will be reported in chapter 8.

Study 3: Children's Books and Psychological Distance

There is now substantial evidence that cultural products both reflect cultural differences and affect cultural orientations (Morling and Lamoreaux 2008). For example, Tsai et al. (2007) found that popular U.S. storybooks had illustrations that were more likely to depict excited (versus calm) states than Taiwanese storybooks. They also found across cultures that exposing preschoolers to exciting (versus calm) storybooks altered children's preferences for exciting activities and their perceptions of happiness.

Study 3 examined spatial properties of pictures in children's books that either are or are not authored and illustrated by Native Americans. We wanted to see if the differences in distance found in the first two studies are also reflected in illustrations in children's books. Specifically, we focused on children's books that included animals as part of the story. We used two closely related measures of subjective or psychological distance: (1) rater judgments of how subjectively close an illustration was to the reader (very close, close, medium, distant, and far away, converted into a five-point distance scale); and (2) type of "camera shot" or perspective (close up, medium, wide angle, panoramic).

Materials

The children's books were selected according to multiple criteria. First, the books had to be targeted at four- to eight-year-old children; second, they had to include animals; and third, we avoided special-occasion or seasonal books (e.g., books about holidays). We also avoided having more than two books by any given author. For the non-Native-authored and illustrated books we first sampled the sixty highest-selling children's books listed on Amazon.com and ended up with a list of forty-four books meeting our criteria that were to be coded. For the Native-authored and illustrated books, we went to the website of a Native-operated literacy organization, Oyate.com, and sampled forty-two of their recommended books. We verified the Native and non-Native reported status of authors and illustrators.

Coding and Results

We enlisted one Native and one non-Native research assistant to perform the coding. The research assistants were blind to the main hypotheses, but it was not feasible to conceal whether illustrations were done by Native versus non-Native artists. Each coder scored most or all of the books. Intercoder reliability was satisfactory, and disagreements were resolved by discussion. The effects we report are large enough that they would be reliable if performed on either coder's ratings.

Our coding scheme can be applied at both the level of individual illustrations and the level of books (does the book have at least one "close-up" picture?). To establish the robustness of our results we focus on book-level codes, though the effects are larger at the level of illustrations.

Native books were reliably more likely to include illustrations that provided a "close-up" perspective (93 versus 75 percent) and reliably more likely to have a picture rated as "very close" (86 versus 59 percent of books). Both results fit with our analysis of psychological distance. To avoid misleading you, we should tell you that Native books were also more likely to have pictures showing a panoramic (distant) view. This is intended to forewarn you that more is going on than just one main effect of distance. We'll bring out the significance of this finding shortly.

Summary

After our introduction to construal-level theory and its construct of psychological distance, we compared Native American and European American psychological distance from the rest of nature. We reported three different measures of distance (outdoor practices, distance in discourse, illustrations in children's books), all of which point to a cultural difference in subjective proximity to nature.

Although this chapter has been about psychological distance, we see distance as one of many related aspects of potential cultural differences in orientations toward nature or in worldviews. Roughly speaking, "worldview" refers to an organized set of beliefs concerning how the world works. Much of the research on worldview and science education has been particularly concerned with what happens in cases when students' worldviews, derived from the cultures in which they live their everyday lives, clash with the worldview of science as carried out in science instruction. For example, Nancy Allen (1995) has suggested that some Native American students tend to view the world holistically and to reason in terms of cycles, in contrast with science classrooms, which tend to be more linear and reductionist in worldview.

The term "epistemology" is closely related to, if not a synonym for, worldview. Epistemology refers to beliefs about the nature, validity, and scope of knowledge.

One important component of worldviews or epistemologies is how human beings are conceptualized in relation to the rest of nature. "Conceptualized" is a big, ambiguous word, and for now we have focused on psychological distance as the variable of interest. So far we have reported cultural differences in psychological closeness associated with outdoor practices, discourse about nature, and illustrations in children's books. In chapter 8 we describe our studies on correlated cultural differences in perspective taking and sensitivity to context (in the form of ecological orientation). Then, once we have you on board our psychological distance bandwagon, in chapter 9 we introduce some complications.

8   Distance, Perspective Taking, and Ecological Relations

Everything has a role to play.
—Menominee hunters and fishers

We are all related; working in the medicinal garden is making relatives.
—American Indian Center teacher/researcher/community member

In chapter 7 we outlined a number of converging findings pointing to the idea that nature may be "psychologically closer" for Native Americans than for European Americans. In this chapter we provide correlated observations that are consistent with construal-level theory, which claims that distance has important psychological consequences. Among the presumed consequences are greater sensitivity to context or situation and an increased ability to take a different perspective. We begin with the latter.

Perspective Taking

We will offer several sources of evidence suggesting that Native Americans are more likely to take on the perspective of nonhuman components of nature. The first that we will mention is from Sara Unsworth's (2008) dissertation and centers on her interviews with rural Menominee and European American adults. She asked adults to describe their last (or a particularly memorable) encounter with deer (everyone in this part of Wisconsin sees deer, even apart from hunting them).

In addition to coding the content of these stories, Unsworth also video-recorded many of the interviews and then coded spontaneous gestures associated with these stories. The main motivation for doing so was that she had attended two hunter education courses, one on the Menominee reservation and one off it, and noticed that the Menominee instructors (but not the European American instructors) often "became the deer." That is, they would use one hand to signal antlers on their forehead and the other to indicate the tail of a deer. Unsworth found that although adults from both groups were equally likely to gesture, only Menominee adults adopted a deer's perspective in doing so. We should add two cautions. One is that only a minority of adults became the deer, so it is not a cultural convention that every Menominee follows. The other is that Unsworth's (2008) dissertation does not provide information on inter-rater reliability from blind coders. The European American adults tended to "place" the deer in some location using one hand and then indicated motion by moving that hand.

These results were conceptually replicated in the Unsworth et al. (2012) study of five- to seven-year-old Menominee and rural European American children's reasoning about ecological relations. This time the relevant measure is children's spontaneous imitation of an animal's sound. Even though young children's books and parents' play with toddlers may focus on animal sounds ("What does the cow say, Johnny? Moooo! Yes! What does the pig say? Oink!"), not one of the fifteen European American children spontaneously gave an animal's sound. The animals used in the ecological relations task (e.g., bee, deer, bear) are not included in typical parent-child play. Nonetheless, six of seventeen Menominee children engaged in sound mimicry, and this cultural difference was reliable.

It is testimony to Unsworth's observational skills that we even have this measure. Once this sort of practice is called to your attention it is easier to see it. Early in our efforts to create culturally based and community-based science education programs, we noticed that before going outside for some activity our (Native American) teachers often stopped and asked the children to "put on your deer ears" to listen to what is happening outdoors.

A third relevant observation comes from our analyses of illustrations from children's books that are or are not authored and illustrated by Native Americans. Our coding scheme included two codes for "camera shots" that invite the reader to take a character's perspective: over the shoulder and embodied. In an *over-the-shoulder shot* the scene is presented as if one were looking literally over the shoulder of a protagonist, and in an *embodied shot* the viewer sees the scene through the eyes of a protagonist (the latter is often indicated by a cut-off view of the protagonist's arms impinging on the scene). For simplicity and to increase coding reliability we combined these two perspective codes. Native books were substantially more likely to employ over-the-shoulder shots or embodied shots (67 percent of books versus 27 percent) than non-Native books, and, when they did so, commonly presented a nonhuman actor's view.

Further analyses of these same books reveal that Native American illustrators are also reliably more likely to use a variety of viewing angles (e.g., high and low angle in addition to the standard, straight-on view) and, as we mentioned in chapter 7, more likely to present a wide or panoramic view (despite the overall tendency to have more "up close" views). In other words, the Native children's books both invite the reader to take the perspective of an actor and employ devices that encourage multiple perspectives in their stories.

Importance Rankings and Perspective

Many resource conservation and environmental decision-making issues reflect a conflict between human desires and what is best for the health of an ecosystem. Our studies with Menominee and European American hunters and fishers reveal cultural differences in values that are consistent with Menominee outdoorsmen incorporating a nature-centered viewpoint into their personal values. We'll describe one study with Menominee and European American hunters (Ross, Medin, and Cox 2007) in some detail.

Initially we asked a sample of hunters to name the most important plants and animals of the forest. From the resulting list we selected twenty-nine animal and thirty-nine plant kinds. Next, we asked each hunter to indicate his familiarity with each kind ("have heard of the kind," "could recognize one," and "have seen one"). Participants were also asked to rate (on a seven-point scale) the importance of each kind to the forest ("How important is X to the forest?") and to themselves ("How important is X to you?"). Instructions were intentionally ambiguous to keep the rationale for an individual's ratings as unconstrained as possible.

We employed peer nomination ("Who are the expert hunters in your community?") to select our participants.[1] On average, nominated experts had been hunting for over twenty years; the data to be reported are based on seventeen Menominee and fifteen European American hunters. Exploring familiarity (with plants and animals) is important because hunting has a focus on game and we were curious to see whether we would find group differences with respect to familiarity with, as well as the evaluation of, plants and nongame animals.

Importance ratings directly test our hypothesis that European American and Menominee hunters evaluate nature from different epistemological standpoints. If Menominee hunters are more likely to take a nature-centered perspective, we should find higher importance ratings for a greater range of flora and fauna for Menominee than for European American hunters. Epistemological differences should also show up in justifications and in the relation between the ratings for importance to self and importance to the forest. For example, if importance to self is based on personal goals, it may conflict with or be uncorrelated with the rating of importance to the forest. Alternatively, if a hunter values the health of the forest, then there may be a correspondence between the ratings of importance to the forest and importance to the self.

Familiarity

Although there were some minor group differences, familiarity with our sample of plants and animals seemed to be more or less equal for members of the two groups. There was greater familiarity with animals than plants, but most of our experts had at least heard of the vast majority of the plants. (This is in sharp contrast with Northwestern University undergraduates who report not having heard of common Chicago-area trees like hawthorn, catalpa, and cottonwood.) Given that we selected our sample of species based on a free listing of species that participants knew, this is not surprising.

Importance Ratings

Recall that this task was designed to test for effects of different epistemological frameworks. The hypothesis is that Menominee hunters would show higher importance ratings for a greater variety of plants (because everything has a role to play in nature and Menominee approach flora for multiple purposes). First, it is noteworthy that we had to exclude the responses of six Menominee (and only Menominee) for the lack of variance within their ratings. All six gave the highest possible importance ratings to all the species. Although we cannot include these responses in our analyses,[2] these data should not be treated as noise. On the contrary, this response pattern marks the Menominee worldview that "all things are interdependent and consequently, equally important."

Despite these excluded data, Menominee hunters gave reliably higher ratings for plants with respect to importance to the forest (means of M = 5.3 versus EA = 4.5). Essentially the same pattern was observed for ratings of importance to the self. Again, the main effect is statistically reliable (means of M = 5.2 versus EA = 4.1). One challenge in this sort of research is to determine whether the differences observed in ratings reflect use of the scale or real differences in valuation (Does a European American "5" reflect a higher value than a Menominee "6"?). To address this question, we can look at justifications for answers.

First, nine of seventeen Menominee hunters provided justifications in terms of general statements that every plant has a role or part to play and hence is important to the forest. No European American hunter provided this type of justification. Second, with respect to ratings of the importance to self, several Menominee hunters mentioned that if something is important to the forest, then it is important to them. Again, no European American hunter provided this kind of justification.

Another aspect of our group differences is that Menominee hunters view the forest from multiple perspectives and goals and not just as a source of game or timber. Within their justifications for their rating of importance to self, Menominee hunters mentioned more uses or sources of value for both plants and animals than did the European American hunters. There was a reliable difference for use of plant material (including medicinal uses of plants) and for justifications in terms of religious, cultural, or symbolic value (and in the case of animals, clan relevance).

The high importance values reported by the Menominee are just one side of the story. In comparison, European American hunters were more likely to report either that a plant had little use to the forest or that they could not think of any. We suspect that this reflects both a lack of knowledge and a more narrow definition of value.

Importance of Animals to the Forest

The mean ratings of the importance of various animals to the forest allow us to see whether the two groups differ in their focus on game animals. Overall, Menominee hunters consistently gave higher ratings for both importance to the self (M = 4.5 versus EA = 3.7) and importance to the forest (M = 4.9 versus EA = 3.7). We find no difference for the rating of game animals. Menominee, however, rated nongame animals significantly higher than European American hunters. This last result is important on two accounts. First, it further undermines the notion that group differences in ratings might reflect different use of the rating scale. If that were the case, Menominee hunters should be giving higher ratings in both cases. Second, it again supports the hypothesis that, in contrast to European American hunters, the Menominee use multiple perspectives to evaluate animals, hunting being only one of them.

With respect to the importance of animals to the forest, Menominee hunters gave a below-midpoint rating only for the nonnative opossum (compared with low ratings for raccoon, chipmunk, otter, porcupine, turtle, blue jay, robin, and skunk for European American hunters). The idea that everything has a role to play may promote deeper analysis of how a species may help the forest. A good example of this is the description of whether porcupines help or harm the forest. A common response among almost all majority-culture hunters was to note that porcupines are destructive because of their habit of girdling and killing trees. Menominee know about this effect too, yet some gave positive ratings and justified them by explaining that this action opens up light into the forest, which in turn allows smaller plants to grow, which in turn provide ground cover that helps maintain soil moisture.

In many respects our findings on importance ratings are striking. Although both groups were more or less equally familiar with the plants and animals employed, cultural group had a large effect in all ratings. Menominee hunters consistently gave higher overall ratings. Justifications for ratings reveal that group differences derive from abstract principles and a variety of species-specific considerations. The abstract principle that many Menominee expressed is that every kind has a role in the life of the forest. This orientation also carried over into ratings of importance to self; a fair number of Menominee hunters mentioned that if some plant or animal is important to the forest, then it is important to them. In comparison, European American hunters were more likely to use a straightforward utility evaluation. Both groups have a rich understanding of the forest, but overall similarities help to highlight group differences.

Finally, we can add a piece of converging evidence from our studies of Menominee and European American fisherfolk. In one study (Burnett et al. 2005) we asked for goodness-of-example (e.g. "How good an example of the category fish is a sturgeon?") ratings for local fish species. As expected,

Menominee fishers gave higher ratings overall. There were no reliable group differences for game fish or for food fish (e.g., bluegill, sunfish), but Menominee fishers gave reliably higher ratings for what the Wisconsin Department of Natural Resources refers to as "rough fish." Rough fish (e.g., suckers, dogfish, gar, carp), commonly referred to as "garbage fish," are generally considered to be undesirable. Menominee fishers might say of a fish like the gar, "I have no use for them, but they must have some function."

We have also used a feature-listing task ("Can you say a little bit about dogfish?") with fish experts, and European American experts are reliably more likely to describe rough fish in negative terms (Medin et al. 2006). Overall, for both hunters and fishers Menominee participants are reliably more likely to value species in terms of nature-centered roles than European American hunters and fishers.

Context and Ecological Relations

Attention to and Importance of Context

According to construal-level theory, psychological closeness is associated with increasing attention to context. Attention to context can be measured in a variety of ways (Masuda and Nisbett 2006; Nisbett and Masuda 2003), and we will draw on data from the interviews with Menominee and European American parents and grandparents in chapter 9. One of our probes asked participants to tell us about the last time they went fishing or a particularly memorable time when they were fishing. Our dependent variable was how quickly adults "got to the point" by mentioning fish. Our idea was that attention to context would lead Menominee adults to spend more time describing the context before talking about fish.

And that is what we found. The median number of words used before mentioning fish was twenty-seven for European American adults and eighty-three for the Menominee adults, a large and reliable difference (Atran and Medin 2008). The reason we had to use medians rather than means is that several Menominee adults never got around to actually mentioning fish.

Taxonomic Relations

There is marked cross-cultural agreement on the classification of living things, such that plants and animals are grouped according to a hierarchical taxonomy with mutually exclusive groupings of entities at each level (Atran

1993; Berlin 1992). Furthermore, the genus (e.g., trout, oak) level appears to be consistently privileged for both naming (Malt 1995) and inductive inference when generalizing properties attributed to one biological kind (Coley, Medin, and Atran 1997). The privileged level is defined as the most abstract level for which inductive confidence is strong, and only minimal inductive advantage is gained at more subordinate levels. For example, undergraduates (who otherwise could not identify an oak tree) were confident that a property true of white oaks would also be true of red oaks and much less confident that the property in question would be true of red maples.

But there is also considerable variability within these universal patterns in concept formation as a function of both experience with the natural world and cultural salience (two highly related factors). Recall that Northwestern undergraduates seem to know very little about biological kinds such as species of trees. Rosch et al. (1976) found that the level at which urban undergraduates possess the greatest knowledge is the life form (e.g., bird, fish, tree), but for groups that have more direct experience with the natural environment and greater expertise, the basic level corresponds to the genus.

One of the ways to assess how people conceptualize nature is to ask them to sort (names of or pictures of) biological kinds into groups that make sense to them. One can then ask them either to subdivide or to combine these initial groups to produce a hierarchical classification system. The idea behind this procedure is that similar kinds will be placed into the same grouping and dissimilar kinds will tend to be placed into different groups. One can then correlate sorting distance (e.g., things in the same lowest-level category have distance zero, things joined at the next level of abstraction have distance one, and so on) with taxonomic distance (measured the same way), and when one does so one typically finds quite high (e.g., +.70) correlations (Atran and Medin 2008). This suggests that a taxonomic organization is natural for participants.

There are two important observations associated with these findings. One is that rather than saying something like, "Well, I could sort these in lots of different ways, what are you looking for?," participants express little uncertainty and almost always quickly come up with an organization that seems natural for them.[3] The second observation is that a correlation of 0.70 explains about half of the variability (R-squared equals 0.49), leaving open the possibility that other factors may be playing a role in sorting. For example, a sorting system based on land versus air versus aquatic animals may correlate with taxonomic distance because those spatial factors are correlated with taxonomic distance (birds are mainly air creatures, mammals mainly ground creatures, and so on). That brings us to ecological relations.

Ecological Relations

Everything we have covered so far leads to the expectation that if biological kinds are psychologically close and closeness leads to attention to context, then it should also lead to a greater sensitivity to (ecological) relations. If that's what you've been thinking, you won't be disappointed.

In the first study we'll describe, expert Menominee and European American fishers from rural Wisconsin were asked to sort names of local species into categories (Medin et al. 2006) and to explain the basis for their sorting. European American experts tended to sort taxonomically (e.g., these are the bass family, these are minnows and shiners, etc.) or on the basis of goals (e.g., these are large, prestigious game fish, these are fish that are good for children to catch, these are garbage fish, etc.). Many Menominee experts also sorted by goals and taxonomic relations, but about 40 percent of them sorted ecologically according to habitat (these are the fish that are found in cool, fast-moving water, these are found in stagnant ponds, etc.). This latter basis for sorting was rarely seen among the European American experts. In a follow-up study with less expert but equally experienced Menominee and European American fishers (Medin et al. 2002), the European American sample was even more likely to sort by goals and the Menominee sample displayed ecological sorting at the same level as Menominee experts.

We then decided to study ecological reasoning more directly. In our second study with fish experts we selected a subset of twenty-one species that all of the experts were familiar with, and for each of the 210 possible pairs asked whether one fish affected the other or vice versa (e.g., "Does the northern affect the river shiner or the river shiner affect the northern?"). The task was completed in about an hour, so with 210 pairs you can imagine that we moved at a fairly rapid pace.

Generally, Menominee and European American fish experts agreed with each other on the relations present (Medin et al. 2006, Experiment 2). If we look at relations that were mentioned by at least 70 percent of one group, then 85 percent of the time 70 percent or more of the other group also mentioned a relation. But there were also striking differences. Only 1 percent of the time did European American experts mention relations that Menominee experts did not agree were present, but 14 percent of the time Menominee experts agreed on relations that European American experts did not mention. Overall, Menominee experts reported reliably more ecological relations, including reliably more positive relations.

A little detective work suggests that the European American experts were primarily answering in terms of adult fish and that their answers were being influenced by their goals. The 1 percent figure we just gave you may have been cases of overgeneralizations driven by goals. For example, for the pair of river shiner and largemouth bass, European American experts tended to say that largemouth bass eat river shiners; for the same question Menominee experts generally said that they are not found in the same waters (and at least in this part of Wisconsin they are not). For a pair like northern and musky (the larger cousin of the northern), European American experts usually said that a musky will eat a northern. Menominee fishers also mentioned this relation but, in addition, noted that northern fry hatch about two weeks earlier in the spring and that northern fry will eat musky fry. This latter observation was a big hint concerning the basis for our group differences.

The big hint is that, in informal conversations, more than one European American fish expert had mentioned to Medin or Norbert Ross that northern fry hatch earlier than musky fry (and will eat them). Why didn't this knowledge come out on the ecological relations task? It dawned on us that we were looking at cultural differences in knowledge *organization* rather than differences in knowledge per se (after all, these guys were experts and had fished for decades). If your knowledge is organized in terms of habitat or other ecological relations, then it should be easy to retrieve facts about these relations. If instead your knowledge is organized in terms of your goals or taxonomic relations, it should take more time to access ecological knowledge.

We did two follow-up studies to test the knowledge organization hypothesis. In one we asked our experts to sort the local fish by habitat. This is a little tricky as some fish (e.g., northerns) are found in both rivers and lakes. We told our experts that we had multiple name cards for each species and that if a given fish is found in more than one habitat, they should put a name card for that fish in each. We found no group differences in ecological sorting, supporting the idea that both sets of experts knew the habitats for the various fish.

The second study, all modesty aside, was really cool. We again (now nearly two years later) gave the same species relation task, but reduced the number of pairs from 210 to thirty-four, allowing us to move at a very leisurely pace. We made two predictions: (1) the group differences would disappear[4] and (2) the European American fish experts would start to answer relation probes by referring to the entire life cycle of fish.

And, amazingly enough, that's exactly what we found. The earlier probe of thirty-four pairs had yielded twenty-eight relations for Menominee experts versus only seventeen for European American experts; now the figures were thirty-two versus twenty-nine, a tiny, unreliable difference. The shift from seventeen to twenty-nine took the form of European American experts now mentioning relations involving spawn and fry. Overall, our data suggest a large cultural difference in conceptual organization, favoring ecological relations for the Menominee and goals and taxonomic relations for European Americans. And needless to say this difference in relational focus fits nicely with construal-level theory.

Developmental Studies

Once we had noted these cultural differences in adults, a natural follow-up question was whether we would also see cultural differences among children. We already had a hint that we might from a study done by Ross et al. (2003), using an inductive reasoning task. In this method children are taught that some novel biological property is true for one biological kind (e.g., "has andro inside it") and asked whether it might also be true for other biological and nonbiological kinds. Let's call the item children were trained on "the base" and the new items they were tested on "the targets." The idea is that children will tend to generalize to the extent that they see the base and target kinds as similar, and that's what usually happens (Carey 1985). These same studies typically are done with children in and around major research universities, and these schools almost always are located in urban areas.

Ross and colleagues employed this induction task with urban children, but also with rural Menominee and European American children. We'll provide more details on their results in chapter 9; for now, we focus on a single result. Older rural European American children and Menominee children of all ages tended to generalize a property attributed to bees (e.g., "has andro inside") to bears, a biological kind not especially similar to bees. Sometimes children volunteered the basis for their answer by saying that a bee might sting a bear (transferring andro to the bear) or by mentioning that bears eat honey (with the unstated implication that andro was in the honey and would be transmitted by ingestion). In other words, rural children sometimes were employing ecological reasoning, a strategy we have not seen in urban children of any age. (See Medin and Waxman 2007 for a more detailed and technical analysis of children's justifications that reinforces the idea that Menominee children are precocious in their use of ecological reasoning.)

Given the encouraging Ross et al. (2003) observations, we decide to probe rural Menominee and European American children's ecological reasoning more directly (Unsworth et al. 2012). We mentioned this study in describing Menominee children's spontaneous imitation of animals, and now we are ready to give you results on ecological relations. Recall that seventeen five- to seven-year-old Menominee children and fifteen five- to seven-year-old European American children participated in this study. The materials included thirty pairs of pictures of plant and nonhuman animal species situated within their natural habitats. There were fifteen animal-animal pairs (e.g., coyote, rabbit), nine animal-plant pairs (e.g., frog, lily pad), and six plant-plant pairs (moss, birch tree). All species represented in the pictures can be found in the state of Wisconsin. We selected pairs that shared a variety of relations, including taxonomic relations (e.g., eagle and hawk are both birds) and ecological relations (e.g., eagle and hawk both eat small rodents). Many species depicted in the picture pairs shared morphological properties as well (e.g., eagle and hawk both have wings). For purposes of this study ecological relations were defined as responses about relations between the species; they included (a) habitat relations (e.g., woodpeckers live in trees); (b) food chain relations (e.g., chipmunk would eat the berries); and (c) references to other biological needs such as water, sunlight, or soil.

Children in both cultures were more likely to mention habitat relations than either food chain or biological needs. Every child gave habitat responses, which may reflect the fact that habitat information was depicted in the pictures themselves (e.g., moss and a birch tree were both depicted in the forest). But Menominee children gave significantly more food chain responses and more relations involving biological needs than rural European American children.

In summary, the results of this experiment provide direct evidence for cultural differences in children's ecological reasoning; as with adults,

Menominee children were more sensitive to ecological relations than European American children. These developmental studies indicate that an ecological orientation is not a perspective that adults acquire, but instead may reflect the sort of epistemological framework for approaching the rest of nature that may be widespread in terms of both explicit and implicit practices in Native American communities.

Summary

Chapter 7 reviewed a variety of converging measures supporting the idea that there are cultural differences in psychological closeness to nature. The present chapter traced the implications of this difference for perspective taking, attention to context, and sensitivity to ecological relations. If we do a tally, the net picture is pretty impressive with respect to cultural differences; and not just any cultural differences, but precisely those expected as a consequence of a difference in psychological distance.

Differences in perspective taking were revealed in spontaneous gesture and spontaneous sound mimicry, and in illustrations in Native versus non-Native children's books. These perspective differences, in turn, were reflected in importance ratings for plants and animals of the forest as well as goodness-of-example ratings of local fish species. The justifications for these judgments also reveal differences in taking multiple perspectives. We also found evidence of cultural differences in the importance of context revealed in stories about fishing. Finally, we found differences in ecological or relational orientation for both children and adults. These differences are supported by sorting studies, speeded versus unspeeded probes of ecological relations, children's use of ecological relations in reasoning, and response to direct probes concerning ecological relations. Overall our data provide unambiguous support for construal-level theory's predictions concerning the cognitive consequences of psychological distance, and they nicely illuminate a consistent patterning of cultural differences. One could hardly ask for more.

But we do. There are some hints in our data that more than distance and its consequences are in play. In addition, if one analyzes this chunk of research and the cultural practices of the researchers who did it, one might come to some surprising conclusions. It's about time for some complications. Chapter 9 will provide them.

9 Complicating Cultural Models: Limitations of Distance

The world can only be "nature" for a being that does not inhabit it, yet only through inhabiting can the world be constituted, in *relation* to a being, as its environment. (Ingold 2000, 40)

As we forewarned you, we have made some simplifying assumptions in order to show the power of considering only the psychological distance component of orientations toward nature. Also as promised, we're now going to argue that distance by itself won't do all the work we need it to do, so we will have to bring in some other ideas about cultural models and epistemological frameworks. First, however, it is time to make a confession.

From a sociology of science perspective, chapters 7 and 8 have a very Western science orientation. The research framework isolates a single dimension—in this case, distance—and ignores everything else that might be relevant to cultural models and epistemological frameworks. Well, actually, it doesn't ignore these other factors, because many of them may "come along for free."

Consider the following example basketball analogy: Imagine that we were doing a study of basketball-playing skills, focusing on a (cultural) comparison of the Chicago Bulls and the Batavia Bulldogs (a Chicago-area high school basketball team). If these two teams met each other in a serious competition, there is zero doubt that the Bulls would beat the Bulldogs. Now let our cultural analysis kick in. If this theory development follows our previous analysis format, it might focus on height as the key factor in the Bulls' success (the average height on the Bulls is about six feet seven inches or so and for the Bulldogs maybe just about six feet—the latter is a guess; the Batavia Bulldogs website lists the players and notes that they are ranked 15,795th nationally as of early September 2011, but doesn't give players' heights). Furthermore, we could readily identify basketball practices that would be correlated with height, like dunking a basketball and blocking shots. Height would also be correlated with a large number of other basketball-related skills that we could measure. Overall, it's clear that height is critically linked to the "cultural" differences in basketball success that appear in our hypothetical world where these two teams would meet.

So far so good? Not if you know anything about basketball. The first thing to note, if you're not already uncomfortable, is that weight could have worked just as well. The Bulls weigh more than the Bulldogs. Furthermore weight can be conceptually linked with basketball practices like blocking out your opponent when going for a rebound or holding your position when an offensive player tries to back you in so he or she can get closer to the hoop. For that matter we could propose a more encompassing theory that age is the critical factor, age being correlated with both height and weight.

The point is that for a contrast as large as Bulls versus Bulldogs, almost any dimension will do, because there are many correlated dimensions that form a cluster. We don't see anything seriously wrong with this analysis— height and weight are relevant to basketball success differences in a way that they may not be for chess (though the correlation of height and weight with age and the correlation of age with chess skill means that height and weight are not correlationally irrelevant to chess success).

But if you are knowledgeable about basketball you cannot help thinking that this analysis is missing some important dimensions. The Bulls are better passers and better shooters than the Bulldogs, and although some part of this difference may be attributable to size, the key to this success consists of a variety of skills built up over many years of practice. If we could somehow determine how many basketball shots players had taken and how many passes they had thrown and how many rebounds they had collected, it would be quite surprising if the least prolific Bull did not have substantially higher numbers in each category than the most prolific Bulldog.

In cultural research more generally, especially in cross-cultural comparisons, one can often find some dimension of difference that brings correlated dimensions with it and allows us to predict a series of related cultural differences. Psychological distance is one such (well-motivated) dimension. But distance may not capture the core of cultural models of nature any more than height captures the core of basketball success.

In this chapter we are going to try to do three things. First, we consider Native American relational epistemologies much more broadly than we have previously and offer additional analyses of Native and non-Native children's books in support of this broader framework. Then we return to the construct of psychological distance and describe some of its concrete limitations. Finally, we describe some cultural developmental studies focused on one facet of cultural epistemologies: the relation of human beings to other animals.

Just a comment on our use of the term "epistemology." In philosophy, epistemology typically is treated as the study of how we know what we know. It may even have a normative component in prescribing strategies that promote having knowledge that is well justified. In other areas like educational science and anthropology, epistemology is treated from a descriptive perspective where the focus is on studying knowledge-capturing processes and practices. In anthropology in particular, researchers have suggested that cultures may differ in epistemological orientations, with correlated differences in understanding our world.

Native American Relational Epistemologies

Anything we write about relational epistemologies will be at once too little and too much. There is a substantial literature on relational epistemologies (e.g., Anderson 1996; Deloria 1998; Kawagley 1995; Nadasdy 2003; Pierotti 2011; Pinxten, van Dooren, and Harvey 1983), and we cannot hope to provide more than a glimmer (*Merriam-Webster* describes "glimmer" as a "subdued, unsteady reflection"). We begin with three quotes:

The *cogito, ergo sum* tells us, "I think, therefore I am." But Native Philosophy tells us, "We are, therefore I am." (Burkhart 2004, 25)

Native science reflects a celebration of renewal. The ultimate aim is not explaining an objectified universe, but rather learning about and understanding responsibilities and relationships and celebrating those that humans establish with the world. Native science is also about mutuality and reciprocity with the natural world, which presupposes a responsibility to care for, sustain and respect the rights of other living things, plants, animals and the place in which you live. (Cajete 2004, 55)

If there is one truly universal Amerindian notion, it is that of an original state of non-differentiation between humans and animals, as described in mythology. . . . The relations between the human species and most of what we would call "nature" take on the quality of what we would term "social relations." (Viveiros de Castro 2004, 464, 465)

Imagine that you are pursuing a career as a teacher and have completed your practice teaching and have been lucky enough to secure a job at an elementary school in a modest-sized town. Of course you will need to be able to handle the content part of teaching and to interact effectively with your students, but a vitally important part of your position will be building relationships in your community. You may not mind being asked to take tickets at your school's basketball game because you'll get to see and chat with a lot of parents. In fact, building relationships may well be a central focus of your efforts.

The same holds for Native relational epistemologies. For example, Cajete (1999) argues that indigenous thought is foundationally based on constructions and meanings of relationships. Some scholars suggest that conceptualizing nature in terms of social relations does not represent an application or transfer of the social world to the natural world so much as the absence of a distinction between the two. Native teachers involved in our project to implement culturally based science education spontaneously describe learning about plants and animals as "making relatives."

Yet another factor motivating us to attend more to the relational side of things is the failure of another analysis we attempted to do with Native- and non-Native-authored children's books. For this iteration our goal was to code the books for moral content. Our subjective impression was that the Native-authored children's books were full of moral substance. Nonetheless our attempts to develop a coding system to capture moral teaching were utter failures; they felt very much like the proverbial effort to pound a round peg into a square hole. Put differently, the Native-authored books seemed to deal with living in (proper) relationship(s), and it wasn't obvious that one could isolate any special subset of this relational complex and call it "morality."

Living in Relation

Of course, our impression that the Native-authored children's books were about relationships did not come with a set of instructions on how one might code for them. Before turning to our new analyses of children's books, we want to make some rough-and-ready distinctions about what a relational epistemology might entail (bearing in mind that there may be many distinct systems that might fall into the category of relational epistemology). On one broad level one can ask what is being related to what.

This is like a list of characters in a play, the dramatis personae if you will. Although this may not seem to be central, we think it is, as cultures may differ dramatically in what they consider relevant and worthy of attention.

In some related research in Wisconsin we have presented ethical and environmental decision-making dilemmas involving tradeoffs such as the following: "Low water caused by a drought threatens twenty species of fish. You can save these species by opening a dam, but opening the dam would cause the extinction of ten different species of fish. Would you open the dam?" Native American (Menominee) and most European American informants were engaged by these dilemmas and often sought out more information (e.g., "What species of fish are in danger?" "What species will die if I open the dam?"). But in a few cases with fundamentalist Christian European Americans, the response was just to laugh. By way of explanation they would say something like "they're just fish." Before jumping to conclusions about lack of caring, keep in mind that a number of the other dilemmas we gave involved human lives and some fundamentalists no doubt found the contrast to be ironic. Still, the fact that no one else laughed suggests a group difference in what is seen as relevant.

The second broad issue is the nature of the relation between and among the entities that are being linked. For example, the relation could be one of reciprocity or it could be asymmetrical. In our work in Guatemala, Itza' Maya saw relations between many species of plants and animals to be reciprocal and positive, but Ladino informants denied that animals help plants and reported that the only positive relations were plants helping animals.

The third broad issue is what larger context and dynamics these relations operate within. This systems-level focus might consider whether there are expressed or implied emergent properties that go beyond sets of pairwise relations. For example, cultural models may differ in the depth of causal chains that are assumed, analyzed, or inferred. Recall, for example, that many Menominee hunters and fishers assumed that "everything has a role to play," even if they had no specific idea about what that role might be.

In a minute we will describe some further analyses of children's books, this time aiming to get at relational construals in a number of ways. For now, however, we want to show that relational epistemologies (and differences in epistemologies) have implications both for how science gets done and for science education practices. For present purposes we propose to sharpen the lens on relational construals through attentional habits, hypothesis generation, and explanatory structures, three practices central to science and science education.[1] Our comments on attention and observation draw on the dissertation work of a Native scholar on our research team, Ananda Marin, and we thank her for her contributions to this section.

Culture and Attention

Several scholars have argued that the ability to attend to objects and events is culturally acquired through the negotiation of attentional directives and participation in routine activities (Cook 1999; Garrett and Baquedano-López 2002; Yont, Snow, and Vernon-Feagans 2003). Attentional habits develop in social contexts, vary across cultures, and vary beyond those practices typically found in middle-class European American communities (Chavajay and Rogoff 1999; Correa-Chávez, Rogoff, and Mejia Arauz 2005; Orellana and D'warte 2010).

Work by Nisbett and colleagues shows that individuals from Eastern cultures tend to direct attention to the field while individuals from Western cultures often direct attention to an object (Nisbett et al. 2001). Similarly, our prior work provides evidence pointing to cultural variation in the kind of relations (e.g., ecological, taxonomic, utilitarian, food chain, biological kind-natural inanimate) that young children attend to (e.g., Unsworth et al. 2012).

The study of scientific observation is one line of work where focus has been directed toward attentional habits. Observation, or systemically seeing and recording phenomena in the world and using these records to build knowledge, is identified as a component of inquiry in reform approaches for science teaching (Gelman et al. 2010; Gelman and Brenneman 2004). Interestingly although observation almost always gets mentioned, it is implicitly treated as atheoretical, acultural, and unproblematic (National Research Council 2007, 2009).

But observation is more than just seeing (see Daston and Lunbeck 2011 for a historical review); it is often driven by some specific theory (Kuhn 1962) and sometimes is used to confirm theories. Attention and observation are directly linked to hypothesis generation in the transparent sense that if you don't notice something, it will neither be the target nor the agent in your thoughts. Observing involves the coordination of attention habits, domain knowledge, and theory (Eberbach and Crowley 2009; Haury 2002), but we know little about the cultural aspects of this process. Correa-Chávez,

Rogoff, and Mejia Arauz (2005) maintain that attention research has been biased toward "focused" versus "unfocused attention," an indication of most researchers' European American, middle-class backgrounds.

Explanation

One aspect of scientific sense making is constructing explanations of phenomena. Researchers interested in science education reform and scientific practices often use Toulmin, Rieke, and Janik (1984) components of argumentation and explanation to assess students' scientific reasoning in classroom settings (Driver, Newton, and Osborne 2000; Erduran, Simon, and Osborne 2004; Lee and Lin 2005; Sadler and Fowler 2006). Toulmin's system is based on informal argumentation in which two parties disagree and then try to defend their positions. But this assumption may not hold for cooperative teams working toward a common goal and may have limited applicability to teamwork associated with science and science education (Marin and Bang in revision; see also Bloome 2005; Kuhn Berland 2008). And Toulmin's system may reflect a singular cultural orientation that may not hold where relational epistemologies are important. For example, consider the roles of storytelling within indigenous communities and their potential impact on science sense making. Many traditional stories include relational accounts of place across time and strata. Creation stories explain phenomena in the natural world and metaphorically describe ecological relationships (Cajete 1999; Settee 2000; Pierotti 2011).

Cajete describes Native science as "essentially a story, an explanation of the how and why of the things of nature and the nature of things" (Cajete 1999, 13). The role of storytelling in sense making also has been taken up in scholarly work outside of indigenous communities (e.g., Goodwin 1984). Ochs et al. (1992) propose that "everyday collaborative storytelling is an experience in dialectic theory-building wherein interlocutors jointly construct, critique, and reconstruct theories of mundane events" (38). We think that Ochs and colleagues' framework may be a better description of how research teams work in indigenous communities.

Our consideration of attention and explanation aims to make three related points. First, indigenous relational epistemologies are not simply abstract stances or principles, but also are embedded in practices that determine the expression of basic cognitive processes like observation and sense making. Second, observation and sense making are at once cultural as well as key components of science-related practices. Third, these sorts of considerations cannot be captured by psychological distance by itself.

Children's Books: Text Analysis

In our analyses of Native- and non-Native-authored children's books, for this iteration we're focusing not on the illustrations but the text. For this iteration of our analysis, we entered the words from forty-four Native-authored and forty-four non-Native-authored children's books into searchable text files. (By the way, if "non-Native" is not a fluent category for you, imagine how fluent "nonwhite" is if you are not white.) In each case the forty-four were a random subset of our original pool of books. (See Dehghani et al. 2013 for further details on the books, coding, and analyses.)

The first analysis we will describe used the Pennebaker et al. (2007) LIWC (Linguistic Inquiry Word Count) application that is available on line. LIWC employs about sixty output categories that reflect linguistic and psychological processes. The application includes a "dictionary" of the assignment of words and word stems to these categories. For example, "we," "let's," "our," "ourselves," and "us" are some of the words that would be assigned to the personal pronoun category "we." Other categories correspond to tense, various grammatical categories, affect, time, quantities, some noun categories, and even forms of punctuation.

A key advantage of LIWC is that it is easy to use and the categories have already been established, so our own team's biases cannot affect the categorization scheme. But this advantage is also a disadvantage, precisely because the categories and the dictionary words assigned to them have not been developed with cultural epistemologies in mind. The categories also would not necessarily translate directly into another language. Spanish requires two forms of past tense where English typically has only one.

We don't want to exhaust your patience, so we won't discuss each of the LIWC categories and their relevance or lack thereof for epistemologies. Some of the categories like "relativity" sound appropriate for relationality, but this category is not only very broad (comprised of some 638 words in the LIWC dictionary), but also is focused on space, time, and motion.

Whenever we could make a straightforward connection between LIWC categories and epistemological orientations, we relied on LIWC. For example, our studies of indigenous scholarship suggest that Native texts should be more likely to establish context, and two ways of doing so are to give background information, which requires the use of *Past tense* (an LIWC category), and to describe relations by using (primarily spatial) *Prepositions*. More speculatively, we thought that the Native propensity for linking events might be reflected in the use of the LIWC *Cause* category. Hence, several of the LIWC categories were relevant and appropriate.

LIWC also has a *Human* category, and we had some reason to think that Native texts would make more use of it. Other LIWC categories that stood out for us as getting at relationality are the following: *We, You, Social* (though this category is also fairly broad and diffuse), *Family*, and *Home*.

The results generally matched our expectations. Native-authored books were reliably more likely to use *Past tense*, more likely to employ (spatial) *Prepositions*, and more likely to have words in the *Cause* category. They were also more likely to have "we-words" and much less likely to have (distancing) "you-words." They were reliably more likely to have words in the *Social* category and the *Family* category. Surprisingly, the non-Native-authored books were more likely to have words falling into the *Home* category. A closer look, however, suggests that a *Home* is not necessarily a home, as it includes words like *pet, sofa,* and *fridge*.

Overall, this initial analysis yielded promising results. There also were a number of other differences that bear further examination. For example, the Native-authored books had more than twice as many instances of words falling into the *They* category. However, to interpret this difference one would need to examine the surrounding context to see how "they" is used (e.g., "They say that when the geese fly low in the fall, we will have a harsh winter"). Note that we could probably disqualify our *You* category on the same grounds (e.g., "You know, it's getting late" doesn't seem to be distancing).

A second, equally important analysis relies on new word categories that we created. This is a more bottom-up approach and involves building a different dictionary tailored to a relational framework. First, consider what is worthy of attention. We predicted that Native books would be more likely to include words corresponding to *Natural inanimates* (e.g., *fire, ice, river, rock, ground, beach, sun, moon, wind*) and *Cycles* and *Seasons* (*birth, death, winter, spring*). They should be also more likely to name nonhuman biological kinds (e.g., *tree, cedar, pine buffalo, coyote, deer, eagle, spider, fish, salmon, turtle*), and when they do so to mention native rather than exotic species.

All of these predictions were reliably supported. Furthermore, when animals and plants were mentioned in Native books, they were given more specific category labels (*trout* rather than *fish*).

We also analyzed kin terms and separated them as primary (so-called nuclear family terms like *father, mother, brother, sister*) versus second-order (no value attribution intended) or extended family (kin terms like *grandmother, uncle*). The two sets of books did not differ on the frequency of primary kin terms, but Native books used extended family terms reliably more often than non-Native books.

Non-Native books were more likely to include words corresponding to human artifacts such as *bed, boat, book, cap, house, road,* and *town*. They also had higher frequency for action verbs like *eat, call, fly/flew,* and *grow/grew* (not to mention the quintessential marker for action, *exclamation point*).

The next step in our analyses could include what one might call relation-process verbs like *tell, visit, share,* and *become*. For the moment this remains a promissory note.

There's also a lot more work to be done to capture the complexity of what we gloss as "living in relation." Consider, for example, one of our favorite children's books, *Yetsa's Sweater* by Sylvia Olsen (2006), which describes the Cowichan sweaters knitted by Coastal Salish women. Yetsa and her mother go to see Yetsa's grandmother. They gather and clean fleece (this includes Yetsa taking "sheep poop" out of it), tease the wool, and watch the grandmother spin it and then knit the sweater with its characteristic whales, waves, wooly clouds, and blackberries (while Yetsa eats fresh bread with blackberry jam on it). The grandmother says to Yetsa that the sweater tells a story about her family—the flowers are there because her mother loves her garden and the salmon symbolizes her father's love of fishing. It literally seems as if everything is connected with everything else and the sweater is far more than a sweater. We need a coding scheme that captures this network of interrelationships.

Concrete Problems with Distance as a Proxy for Cultural Models

We have already alluded to the difficulty of capturing the generalized Menominee notion that everything has a role to play. The notion in itself is abstract and should be associated with a distant level of construal, and distancing was typical of European Americans, not Native Americans.

Another complication is that distance need not be symmetrical. If A is uphill from B, then the psychological distance from B to A may be greater than the distance from A to B. In addition, we prefer to compare the variant to the ideal or standard rather than the standard to the variant. For example, ninety-nine may be more similar to one hundred than one hundred is to ninety-nine, and we say that the teacher met the President of the United States rather than the President of the United States met the teacher (Gleitman et al. 1996; Tversky 1977). Even when there is no clear standard or ideal, saying that A is like B (e.g., wolves are like dogs) means something different from saying B is like A (e.g., dogs are like wolves; Bowdle and Gentner 1997, 2005; Medin, Goldstone, and Gentner 1993). In our example of dogs and wolves, one could also have a nondirectional comparison by simply stating that dogs and wolves are similar and that may bring different things to mind than either directional comparison. The overall construct of psychological distance assumes that distance is symmetrical and, as we will see, this may miss some important distinctions.

For example, questions probing other key aspects of how people conceptualize human beings in relation to the rest of nature include: (1) Is there an ideal or standard? and (2) Are comparisons nondirectional or directional, and if directional, what is the direction of comparison? Furthermore, having a clan system based on animals (e.g., the major Menominee clans are bear, eagle, moose, wolf, and crane) may carry the implicit assumption that humans and other animals are similar (nondirectional comparison) or that humans are like other animals (directional comparison). (Not to mention that scholars like Eduardo Vivieras de Castro and Tim Ingold describe a number of other ways in which indigenous people may conceptualize animals, humans, and their interrelationships.)

The Menominee origin story has people emerging from the bear,[2] so one might even consider the bear as an ideal or standard. Now consider a typical animated movie (e.g., the Dreamworks film *Over the Fence*) where animals wear clothes, drive cars, and so on. These movies have the implicit message that animals are like humans, a clear directional comparison, presumably with humans as the standard. Note, however, that although comparisons may invite attributing base characteristics to the target of a comparison, comparisons can "backfire" if the contrasting differences loom larger than the similarities. K-2 in the Himalayas is like Mount Everest, but the typical hill outside of Santa Fe may suffer by the comparison. In the same manner it isn't clear whether anthropomorphic movies promote or undermine similarities of (nonhuman) animals to humans.

Psychological closeness may increase attention to context and situation, but this may not be sufficient, in itself, to encourage an ecological orientation or systems-level thinking (see also chapter 9 of Berkes 2008). In particular, one can be psychologically close to the biological world and still adopt a markedly anthropocentric orientation (Epley, Waytz, and Cacioppo 2007; Waytz, Cacioppo, and Epley 2010). We suggest that Native American communities' practices—both direct and indirect—encourage taking multiple perspectives on nature, and promote psychological closeness to it, but are not anthropocentric.

Much of the work on psychological distance has contrasted situational with dispositional interpretations of human social behavior but has not elaborated on a relational orientation more broadly, even for human social behavior. Native American epistemological orientations elaborate psychological closeness by focusing on principles of "living in relation," where the relations include not only plants and animals but also natural kinds (e.g., rocks, water).

There is a great deal more that can be said about the particulars of relational epistemologies, including such things as spiritual entities, grandfather rocks, and the like. In the remainder of this chapter, however, we will focus on some cultural developmental studies looking at only the relation between human beings and other animals. This work will address the claim that children's biological cognition includes a mandatory stage of anthropocentrism.

Is a Human-Centered Biology Hardwired?

A general trend in the cognitive sciences has been a shift from viewing human beings as general-purpose thinking and learning systems to seeing them as adaptive and adapted organisms whose cognitive mechanisms are specialized and contextualized to their particular environment (Cosmides, Tooby, and Barkow 1992). In this view, learning may be guided by certain skeletal principles, constraints, and (possibly innate) assumptions about the world (e.g., see Gelman 1990; Keil 1981; Kellman and Spelke 1983; Spelke 1990). In an important book, Carey (1985) developed a view of knowledge acquisition as built on framework theories that entail ontological commitments in the service of a causal understanding of real-world phenomena.

Put more simply, different causal principles may be in play in different domains. For example, the (physical) laws that apply when a bat hits a baseball may be different from those that apply when a parent tries to get her child to "hit the books." A test used to determine whether two concepts refer to ontologically distinct entities is that these concepts are embedded within different causal explanatory frameworks (Inagaki and Hatano 1993; Solomon et al. 1996). Candidates for distinct domains are physical processes and events (naïve or folk physics), biological processes and events (naïve or folk biology), and psychological events and processes (naïve or folk psychology).

Developmental psychologists have done a number of very clever studies with infants that reveal that very early on babies understand certain physical principles (e.g., Carey 2009; Spelke, Phillips, and Woodward 1995). Studies of infant perception and causal understanding suggest that many of the same principles underlie both adults' and children's concepts of objects (e.g., Baillargeon 1994; Spelke et al. 1992). For example, common motion appears to be a key determinant of four-month-old infants' notion of an object, and infants also act as if they know that two objects cannot occupy the same space at the same time. (It is not too much of an exaggeration to imagine babies as patiently waiting for developmental researchers to come up with tasks so that infants can show off just how much they already know.)

One of the most contested domain distinctions, and one that has generated a great deal of research, is that between psychology and biology (see Carey 2009; Herrmann, Waxman, and Medin 2010; Medin, Lynch, and Solomon 2000 for reviews). For U.S. adults who may subscribe to a dualism between mind and body, psychology and biology are distinct domains and have distinct causal principles. Eating a candy bar can give someone instant energy but it will not make them a sweeter person. Carey (1985) argued that (young) children do not distinguish between psychology and biology, but rather that biology is initially understood in terms of folk psychology. On her view, naïve biology emerges as a distinct domain only in older children.

That's a pretty strong claim and Carey (1985) offered some striking evidence to support it. There are two steps to her argument. The first is to claim that while human beings may not be the prototypical animal, they are the premier psychological beings. The second step is to show that children's biological reasoning is organized around human beings as the paragon or prototype. From this one would have a good argument that children's biological reasoning is organized in terms of a naïve psychology. On this view, the distance between humans and other animals is not symmetrical, but rather animals are compared to humans rather than vice versa.

The strongest evidence for a human-centered stance in young children's biological reasoning comes from Carey's own pioneering research (Carey 1985). In an inductive reasoning task involving children (ranging from four to ten years of age) and adults from Boston, participants were introduced to a novel biological property (e.g., "has an omentum"), taught that this property is true of one biological kind (either a human, dog, or bee), and then a few days later asked whether other entities might have this property.

Carey reported dramatic developmental changes in inductive generalizations. First consider the data from the youngest children. If the novel property had been introduced as true of a human, four-year-olds generalized, or projected, that property broadly onto other biological kinds as a function of their similarity to humans. But if the very same property was introduced in conjunction with a nonhuman animal (dog or bee), four-year-olds made relatively few generalizations to other animals. This produced a pattern of generalization that violates intuitive notions of similarity. For example, four-year-olds generalized more from human to bug (stinkoo) than from bee to bug. Overall, Carey (1985) provided two strong indices of anthropocentric reasoning in young children's judgments: (1) projections from humans to other animals were stronger than projections from dog or bee; and (2) there were strong asymmetries in projections to and from humans (e.g., inferences from human to dog were stronger than from dog to human).

In contrast, older children and adults showed no indications of anthropocentric reasoning. Instead they tended to generalize novel biological properties broadly from one biological kind to another, whether the property had been introduced as a property of a human or nonhuman (dog, bee) animal. Moreover, unlike the four-year-old children, their tendency to generalize a novel property was a function of the (intuitive) similarity of the base kind to target kinds (e.g., projections from either a dog or human base led to more generalization to other mammals than to invertebrates or insects). This pattern of induction suggests that for older children and adults, reasoning about the biological world is organized around a concept of animal that includes both human and nonhuman animals.

Carey (Carey 1985; Carey and Spelke 1994) has argued forcefully from these data that young children hold a qualitatively different (and incommensurate) understanding of biological phenomena from that of adults, and that development within the domain of biological knowledge entails a fundamental conceptual change. Carey (1985) entitled her book *Conceptual Change in Childhood* because her data suggested that children begin with a human-centered, psychological understanding of biology and later must reorganize their conceptual system to reflect the understanding that, biologically speaking, humans are one kind among many. More precisely, her claim is that young children view the biological world from the perspective of a naïve psychology, a perspective that must subsequently be overturned as children acquire the mature perspective of a naïve biology.

Carey's provocative claims stimulated a great deal of subsequent research to which we cannot do justice. Some of the subsequent research showed that young children have understandings of distinctively biological mechanisms such as growth (Hickling and Gelman 1995) and inheritance (e.g., Hirschfeld and Gelman 1994; see also Gelman 2003). One intriguing suggestion offered and supported by Inagaki and Hatano is that young children have a distinctively biological framework theory based on the principle of vitalistic energy (Hatano and Inagaki 2000; Inagaki and Hatano 2002). They proposed that cultural models espoused within a community shape children's biological reasoning. Their studies revealed that five- to eight-year-old Japanese children understand many bodily processes in terms of vitalism—a causal model that is pervasive in Japan and that relies on the distinctly biological concept of energy. It remains to be seen whether this is a specific cultural notion or whether biological notions involving energy might be more widespread. We'll take up this notion of cultural models and biology again after a ten-year detour, one that was not entirely wasted even if we were on the wrong path.

Expertise

In the mid- to late 1990s Medin teamed with cognitive anthropologist Scott Atran and a cadre of bright graduate students and postdocs to explore the role of culture and expertise in people's understanding of biology. Our interest in expertise was driven by two main factors. One consisted of close parallels between Itza' Maya elders and U.S. biological experts (Bailenson et al. 2002; López et al. 1997; Medin et al. 1997; Proffitt, Coley, and Medin 2000). The other was corresponding evidence of "devolution" or loss of biological knowledge in technologically saturated cultures such as the United States (e.g., Wolff, Medin, and Pankratz 1999).

It didn't take a genius to come up with the idea that Carey's results reflect urban children having a lack of intimate contact with nature. (For an in-depth discussion of this issue see Louv's 2008 book *Last Child in the Woods*, which argues that children are now suffering from a "nature deficit disorder.") We soon began to employ Carey's inductive reasoning task with rural children and children who presumably have "more intimate contact with nature."[3] When we did so, we did not observe that four- to five-year-olds engaged in the sort of human-centered reasoning that Carey had noted (e.g., Atran et al. 2001; Ross et al. 2003).

Meanwhile an ingenious study by Inagaki and Hatano also pointed to the importance of experience and expertise. Consistent with this idea, Inagaki and Hatano (Inagaki 1990; Inagaki and Hatano 2002) found that urban children raised in Tokyo who were closely involved with raising goldfish generalized biological facts to kinds similar to humans *and* to kinds similar to goldfish. This suggests that the relative advantage for humans over nonhuman animals as bases for induction derives from children's greater willingness to generalize from a familiar base than from an unfamiliar base. Although they did not use Carey's induction task, the anthropocentric pattern produced by urban Japanese children who did *not* raise goldfish converged well with Carey's (1985) results. But the full pattern of results points to a different interpretation—urban children's propensity to view humans as a privileged base may be driven by the fact that humans are the only biological kind that they know much about.

Medin recalls chatting with Susan Carey about the importance of expertise and she offered three responses, all good points: (1) young children's experience with humans (perhaps universally) outweighs their experience with nonhuman animals, and this experience may lead them to privilege humans as an inductive base; (2) the results with rural populations could just mean that rural children get the relevant experience for conceptual change sooner than urban children (that is, maybe all children pass through a human-centered stage but rural children do it sooner); and, by the way,

(3) no one had used a procedure close enough to hers to convincingly show a different pattern of results.

Carey's third point was particularly troubling. Ross et al. (2003) had included young urban children, and although the young urban children showed a pattern that differed from that of rural children, the pattern also differed from Carey's original results.

There's a sense in which we were guilty of cutting corners. Carey's procedure involved teaching a child about only one base and then bringing them back a few days later for generalization tests. That's easier to do when your participants are local. For cross-cultural research requiring travel, it is tempting to be more efficient. We tested for generalization right after training and, after using one base and one novel biological property, we went on to present another base biological kind and a new property, following by a new set of generalization tests, and so on. In fact, Ross and colleagues ended up using five bases, five novel biological properties, and five generalization tests, all in one session.

We didn't think these differences mattered, but the possibility that they did nagged at us. Eventually, we decided that we needed to be able to reproduce Carey's human-centered reasoning results with four- to five-year-old urban children before trying to interpret results from other samples.

And it turned out that Carey was right about her methodological point. Please bear with us for some technical details. We now know that a key procedural variable is whether children are trained on just one base versus multiple bases and whether human appears as the first base or later. These order effects take the following form: young children's tendency to generalize a novel property from a human base to the other animal targets is considerably stronger when the human serves as their first base rather than a later one. Moreover, this order effect is evident across communities, including urban children raised in Indonesia and the United States (Anggoro, Medin, and Waxman 2010). This order effect raises the possibility that anthropocentric reasoning would have been observed if these studies had instead varied base between participants where children see only a single base and order effects are ruled out.

In a follow-up study with urban children, rural European American children, and rural Menominee children, we taught all of the children only about a single base and gave the generalization test a day or two later. Here's what we found (see Medin et al. 2010; Waxman, Medin, and Ross 2007).

The first key result is that we replicated Carey's (1985) pattern of human-centered reasoning for the urban four- to five-year-olds. These young children showed greater generalization for a human base than for a dog base, and they also showed greater generalization from human to dog than from dog to human. In contrast to their urban counterparts, young rural European American children generalized more from a dog base than from a human base. Interestingly, however, they showed greater generalization from a human base to a dog target than from a dog base to a human target. At a minimum this shows that these two measures of human-centered reasoning are separable.

Somewhat surprisingly, like their urban counterparts, four- to five-year-old Menominee children favored the human over the dog as a base when generalizing a novel property to other animals. In part, this may reflect the cultural significance of bear: generalizations from human to bear are especially strong (86 percent) for four- to five-year-old Menominee children, as compared to the urban (67 percent) and rural European American (52 percent) children. But in contrast with urban children, young Menominee children show no evidence of human-dog asymmetries.

In summary, we followed Carey's method with enough fidelity that we were able to replicate her finding of human-centered reasoning in four- to five-year-old urban children. With worries about procedure more or less out of the way, we found that neither rural European American children nor rural Menominee samples demonstrated Carey's two markers of anthropocentrism (human-animal asymmetries and humans as a more effective base than animals).

This pattern of results has two key implications. First, our findings document that the human-centered pattern reported by Carey for young children is far from universal. Second, we establish that two signatures of anthropocentric reasoning in Carey's account—generalization and asymmetries—are in fact distinct (see also Medin and Waxman 2007). This is important because it reveals that these two measures do not necessarily tap into a single underlying model or construal of biological phenomena.

To be sure, each of the rural groups showed one of the markers, and follow-up studies will be needed to probe these findings in greater detail. For Menominee children we would like to examine both the role of clan animals in generalization and the potential influence of ecological reasoning. Young rural Menominee children generalized less from the dog base to other animals than did rural European American children in the study we just described, but Ross et al. (2003) found that young Menominee children generalized substantially more than young rural European American from a wolf base to other animal targets. Wolf differs from dog both in being native and in being a Menominee clan animal.

That takes care of Carey's third point but leaves her first two points intact. How do we know that our rural samples have not gone through the stage of a human-centered biology, but just did it sooner than urban children? The obvious way to address this question is to run three- to four-year-old rural children on the induction task. That's a nice idea but there's a problem—for a task like this, four years old is about as young as you can go and still get meaningful data. Younger children will answer your questions but they may say "no" or "yes" to every probe.

Fortunately Patricia Herrmann, then in our lab (who also benefited from the wise mentorship of developmental psychologist Sandra Waxman), was able to solve this challenge by borrowing a procedure that has been used before with toddlers. One of the problems with the usual procedure is that it is given by an adult, who presumably knows more about biological kinds than the child does. Children may find this arrangement strange since children normally are asking questions of adults. Borrowing from other cognitive developmental work, Herrmann modified the usual method by introducing two puppets, each of which is right some of the time and wrong some of the time (as established in a warm-up task). For the induction task the two puppets disagree about whether some biological kind has the property in question and the child acts as a mediator and casts the decisive vote. With this method and an experimenter who has the skill to establish excellent rapport (as Herrmann does), three-year-olds produce systematic, meaningful data.

Cultural Models Matter

With our lead-up you might expect that the next study we will tell you about was done with three-year-old rural children. But we had two good reasons for starting with urban threes. One is that they were more accessible and we wanted to iron out any procedural wrinkles before presenting this new method to our Wisconsin research assistants. The other was our hunch that a human-centered biology may reflect a cultural model and perhaps one that three-year-olds have yet to acquire. Unlikely as it may seem from the point of view that experience and expertise are the key, we thought that urban three-year-olds would *not* show a human-centered biology. Specifically, if the human-centered reasoning pattern seen in young urban children represents the acquisition of a culturally transmitted anthropocentric model, this should also have developmental implications: it may be the case that urban children younger than four to five years of age, who have received less exposure to the anthropocentric model, would not (yet) favor humans over nonhuman animals in their reasoning.

And that is what we found (Herrmann et al. 2010). Three-year-old urban children responded systematically, generalizing more from a dog base than from a human base and showing no reliable human-dog asymmetries. To make sure that the puppet procedure didn't introduce some artifact, we also ran urban five-year-olds with puppets, and they showed the now-familiar pattern of generalizing more from a human base than a dog base as well as substantial human-dog asymmetries. This pattern has been replicated often enough that we are quite confident of these findings.

We have also used the puppet procedure with four- to five-year-old rural European American and Menominee children just in case it changes the pattern of performance, but have not found either of the markers of a human-centered biology. Unlike the Medin et al. (2010) results, we have not found human-dog asymmetries with four- to five-year-old rural European American children, or that human is a better base than dog for Menominee children.

These results offer unambiguous evidence that the anthropocentric pattern of reasoning observed in urban five-year-old children is not an obligatory initial step in reasoning about the biological world. Instead, the results show that anthropocentrism is an acquired perspective, one that emerges between three and five years of age in American children raised in urban settings.

This interpretation is consistent with evidence from two independent sources. First, beginning in the first year of life, children distinguish animate objects (including human as well as nonhuman animals) from inanimate objects, and invoke this distinction when making predictions about the behavior of objects (e.g., Gelman 2003; Luo, Kaufman, and Baillargeon 2009; Massey and Gelman 1988). We suggest that the three-year-olds in our experiments also invoked this concept of animacy in their reasoning about biological phenomena. Second, these results, coupled with those from four-to five-year-old rural European American and Menominee children, underscore the point that the path of acquisition is importantly shaped by the input that learners receive.

We could now present yet another analysis of children's books, but the differences are so enormous that we see little point in providing numbers. Animals in children's books by non-Native authors are overwhelmingly anthropomorphized, with animals wearing clothes, driving cars, living in houses, and so on. Native-authored children's books hardly ever depict animals this way. Sometimes the animals talk with each other (in English), but we would argue that this reflects sentiments about communication, not anthropomorphism.

Modes of Construal and Priming

In urban technologically saturated communities, where direct contact with nonhuman animals is relatively limited and where images of nonhuman animals in children's books, discourse, and media often take an anthropocentric cast, young children encounter considerable support (intended or not) for an anthropocentric understanding of the relation between human and nonhuman animals. Of course, inputs are not necessarily consistent with each other and may invite or take advantage of multiple perspectives or modes of construal (Keil 1992). For example, Gutheil, Vera, and Keil (1998) did studies with four-year-olds in which they described biological properties of human beings in either a psychological context (e.g., "This person eats because he loves to be at meals with his family and friends. Meals bring the family together to eat and have fun. If he didn't eat he would never see his family all together") or a biological context (e.g., "This person eats because he needs food to live and grow. The food gives him energy to move. If he doesn't eat, he will die"). They then looked to see if these properties would also be attributed to other biological and non-biological kinds. Gutheil and colleagues found a sharp similarity gradient (similarity to human) after the psychological framing and a broader, less human-centered generalization. Interestingly, the data from a "no context" condition were close to those in the psychological condition, suggesting that the human-centered psychological construal is closer to the chronic orientation of their four-year-olds.

Patricia Herrmann has followed up her initial findings by presenting the puppet-based induction task after presenting a three-minute video clip that is either anthropocentrically oriented (e.g., Disney's movie *Robin Hood*, in which all the characters are animals) or more biologically focused. In some cases she uses a neutral, control prime that doesn't feature animals, such as Disney's *Beauty and the Beast*. Her idea is also to see if distinct modes of construal can be primed. There is some evidence that they can. A human-centered prime leads seven-year-old (and three-year-old) urban children to generalize more from a human base; a biological prime leads four- to five-year-old children to generalize more from a dog base.

Anthropocentric images are not confined within city limits, but the consequences of this style of input may be attenuated in rural communities, where alternative perspectives may be more readily available and more explicitly encouraged. Interestingly, presenting an anthropocentric prime to rural seven-year-olds is associated with *less* generalization from a human base. This sort of "backfiring" effect remains to be explored in detail.

These data show unambiguously that patterns of inductive generalization are susceptible to priming effects. So far, however, Herrmann's data are ambiguous with respect to whether priming is associated with distinct modes of construal versus the more conservative possibility that primes work to increase the psychological similarity of one base or the other to different targets. For example, giving seven-year-olds a human-centered prime increases generalization from a human base to other animals but does not appear to reduce generalization from a dog base. In short, an anthropocentric prime does not produce the full five-year-olds' urban signature of humans serving as a better base than dogs.

Summary of Induction Studies

Our initial singular focus on biological expertise or lack thereof got in the way of our seeing the importance of cultural models embodying different relationships between humans and the rest of nature. Furthermore, as anticipated by Epley, Waytz, and Cacioppo (2007) and by Waytz et al. 2010, these differences in models cannot be understood by an appeal to psychological distance. Carey (1985) may have been correct in thinking that biological cognition may involve competing, incommensurable models, but we suggest that these are competing cultural models, not some acultural naïve psychology or naïve biology. We need to understand the dynamics of these various cultural models, which appear to vary both across cultures and within individual minds, depending on the context. And just as obviously, conceptualizing culture differences solely in terms of psychological distance is not going to be sufficient to capture the details of cultural models and biological cognition.

Just a Word about Practical Implications

If our fundamental perspectives on the natural world are shaped by experience, cultural beliefs, and practices, then children from different backgrounds may harbor different perspectives even by the time they enter school. It is therefore important to identify which perspectives children acquire earliest or with least effort, and how these are shaped over development. Addressing this issue and adopting a cross-cultural developmental approach is central not only to theories of conceptual development but also to science education. To design effective science curricula requires us to understand these diverse perspectives and competing cultural models. To obtain more leverage in understanding children's biology and the role of culture in shaping it, it is essential to extend research beyond the induction task to include multiple converging measures (as, for example, in Anggoro 2006).

Summary

The one-sentence summary of this chapter is that the core of the cultural differences we have been exploring goes beyond a singular difference in psychological distance. We do not doubt that psychological distance is a central factor, but we equally believe that much more is needed to capture cultural ideas and processes associated, for example, with indigenous relational epistemologies. This "much more" requires that we begin to link these abstract notions with contexts of practice.

We see the following as a key implication. Neither the practice of science nor its teaching involve bringing some consensus common foundation to the table, at which point different cultures apply different seasonings to give rise to distinctive flavors. Instead cultural factors are at the core of these practices and play out well in advance of anything being brought to the table. On this view, science education will not succeed by bringing in "cultural connections" at the end nor will middle-class, white cultural researchers succeed by going about their business and then asking people from the communities being studied to comment on and approve their work. These strategies may (or may not) be better than nothing, but they seem to miss the point. We'll pursue these ideas and implications further in the next few chapters.

10   The Argument So Far

The paradox of culture is that language, the system used most frequently to describe culture, is by nature poorly adapted to this difficult task. It is too linear, not comprehensive enough, too slow, too limited, too constrained, too unnatural, too much a product of its own evolution, and too artificial. (Hall 1976, 49)

Before turning to the next part of our argument and associated empirical support, we want to summarize where we've been. First, the history of science taught in U.S. schools is a history of Western science, not because the West is where science developed, but rather because those determining what story will be told are Western. In our view, this story is ethnoscience, full of stereotypes and other forms of bias; it is ethnohistory, not history. The development of science is far more global, and for much of its history the West was less technologically developed than the East and the Middle East. Furthermore, throughout the world indigenous sciences made important discoveries about the nature of nature. In brief, the first step to opening up science is to distinguish between science and Western ethnoscience.

The second obstacle to be challenged is the idea that science is based on an objective "scientific method" that is value-free and acultural. This stereotype has it that "true science" is distinct from all other ways of knowing and is something of a culture unto itself. That is, one learns to adopt the practices of science and thereby be able to become objective, much as we learn to wash our hands in order to become free of germs. This stereotype may be good for the image of science, but it is so far removed from reality that it becomes a pernicious lie. When a nuclear physicist tells you that the nuclear reactors in Japan are safe against all foreseeable risks and that the earthquake and resulting tsunami were against all odds, you might want to think about questioning the oddsmakers. Although the practices of science and its social nature help eliminate some sources of bias, they may serve to reinforce other biases, as long as science is owned and operated by a narrow cadre of researchers who may be blind to the potentially self-serving nature of their scientific agenda. Furthermore, when science is largely narrowed to a particular subgroup, it is almost inevitable that this subgroup will engage in "niche construction," creating and reinforcing practices and values that favor that same subgroup and tend to exclude others.

Culture and cultural values affect what problems scientists choose to study and how they choose to study them. For a set of disciplines devoted to truth, a little more truth in advertising seems in order.

The observation that the science that gets done depends on who is asking the questions does not mean that science is fatally biased. Rather it suggests that multiple perspectives are needed for effective science. And there must be power sharing so that there can be "niche construction for all."

The third important step in our argument, but not an absolutely essential one,[1] is to query the idea of the unity of science, realized in terms of a single, correct description of how the universe works. Some who argue for pluralism in science properly note that we shouldn't assume that higher levels of description can be reduced to lower ones (i.e., by stereotype, that physics can explain everything). Our view is that science may be a description or a model of reality (speaking a bit loosely), but that there are an infinite number of accurate (that is, accurate enough to be useful) descriptions of nature that may be variously relevant depending on our goals and values. Again we appeal to the notion of maps—there may be an unlimited number of different kinds of maps, each constrained by what it depicts, and each useful for some purposes but not others. As we have noted, maps are not arbitrary, but the map is not the country, nor is the model of reality reality itself.

Once we grant that science is pluralistic in terms of admitting multiple, accurate (enough) descriptions, the case for broadening the ownership of science becomes even more compelling. Medin recalls years spent at Rockefeller University where graduate students and postdocs from different biochemistry labs were not allowed to talk with each other. Within that community there was the feeling that there was a common goal, a common set of methods, and that it was just a race to determine who would get there first, almost as if science were a treasure hunt with just so much hidden gold to be discovered.

We don't think that the amount of "treasure" to be discovered is limited in that way. This may be especially evident in the social, educational, and behavioral sciences but arguably is intrinsic to all the sciences. The well-worn path may be compelling as the path to some destination, but it may also reflect the lack of diversity in precedents and failure to conceptualize alternative destinations. In short, we endorse a pluralistic stance, one that literally calls out for diversity in terms of who is drawing the maps and who is asking the questions.

The next step in our analysis was to draw on the work of others to show how values affect the practices of science and to illustrate how science reflects the culture (and gender) of its practitioners. Again the question is not about good science versus bad science but rather insights from one way of thinking about things versus insights from other ways of thinking about these same things. Sociologists of science have made a compelling case that science practices are value-laden rather than value-neutral.

All of the preceding review may be seen as background for the research and theory that have engaged Bang and Medin (and our communities) over the past decade or so. The theory side of our project has been to make the case that there are cultural differences in orientations toward nature and that they have clear and compelling cognitive consequences. We have described a range of converging measures that suggest that (the rest of) nature is psychologically closer for Native American children and adults,[2] and the cognitive consequences we have observed are those predicted by Trope and Liberman's (2003) construal-level theory. We also noted that, in addition to distance, there are centrally important cultural differences in what we, and others, describe as a "relational epistemology."

We should also take a slight detour to note that the term "traditional ecological knowledge," or TEK, has also been used to describe indigenous contributions to science. Scholars like Berkes (2008), Henriksen (2009), and Pierotti and Wildcat (2000) use "TEK" to describe an epistemological orientation that is local, relational, and recognizes humans as part of the ecological system, rather than as separate from it. That is very much what we mean by a relational epistemology.

The reason we have avoided using the term "traditional ecological knowledge" is that other (mostly Western) scholars have interpreted TEK in a way that positions it in the past and renders it as primarily of historical interest. Worse yet is the term "ethnoscience," which is used to refer to the study of "non-Western systems of understanding the world." As Hess (1995) notes, one key problem with this term is that all knowledge systems are "culturally rooted," including Western science. He also points out that simply imposing Western terms (e.g., biology, physics, etc.) on other cultural systems may divide their world in artificial and counterintuitive ways.

Although the value of TEK in a wide range of applications is undeniable (e.g., Berkes 2008), the use of "knowledge" invites a view of science as established facts, and "traditional" typically suggests "folk" as a synonym. On this reading, indigenous peoples have made discoveries such as medicinal uses of plants, and modern science can now either confirm these findings as true or dismiss them as "old wives' tales." Although it is better to acknowledge the contributions of Native science to our well-being (e.g., aspirin) than to ignore them, this reading makes Native science stagnant.

In contrast, Native scholars like Cajete, Kawagley, and Wildcat view Native science as dynamic, not a thing of the past, and characterized by a set of practices and orientations more resonant with the "ecological" component of TEK. To avoid the misinterpretations associated with "TEK," we use the term "relational epistemology" as the framework for our studies, and think that the increasingly used term "indigenous knowledge systems" (IKS), out of which TEK is produced, is more promising than TEK by itself.

The observation that our cultural differences seem best described as differences in epistemological orientations toward nature suggests that Native scientists and European American scientists[3] may bring different assets to the table (see also Pierotti 2011). These assets may be used to develop a more diverse and more effective science or they may fail to be utilized or, in some cases, may be actively suppressed. The latter situation may lead to disidentification with science and a sense of alienation. Alternatively, when these assets are recognized and appreciated, Native scholars may flourish.

The remainder of this monograph can be seen as a test of the claims just discussed. We describe our efforts to develop and implement culturally based science curricula and examine their consequences for science learning and identification with science. Our research sites are the Menominee reservation in Wisconsin and areas in and around the American Indian Center of Chicago. There are three related issues that bear mentioning before we turn to this body of work. One is that to call our approach culturally based also entails that it be community-based. This almost surely is the case in general, but the particular history of Indian education in the United States (that is, the education of Native Americans) makes the community-based component foundational. The second factor to note is that cultures are complex systems, and a systems-level orientation is required in order to understand the dynamics of community- and culturally based science education efforts.

One implication of this point is that attention needs to be directed not only to the students engaged in the learning environments but also to teachers, community members, and the research team itself. This point is often acknowledged in educational research where researchers design some intervention and work with teachers to implement it, but here the teachers and students are conceptualized as the system of interest and the researchers themselves remain outside of it. This may reflect the implicit idea that research requires distancing and the researchers must remain outside the system to keep from "biasing it in some way." But for our project this stance won't work—our research team is very much a part of the system of interest, regardless of whatever challenges this fact might pose.

Finally, we should note that a good part of our story reflects the wisdom of hindsight. Just as design research typically goes through multiple iterations, the different components of our project are ongoing and dynamic. To be concrete, the development of our research project has *not* been a linear unfolding from observational studies to cognitive experiments and on to an intervention to test our ideas about culturally based science education. Instead, even as we iteratively design interventions, we continue to collect observations on practices associated with informal learning contexts as well as to perform cognitive studies. Our research strategy is well reflected in figure 10.1, which shows the three domains of inquiry in an ongoing interaction. We see each of these spheres as crucial to our objectives and each as informing the others.

These dynamics aside, the logic of our research is clear. If epistemological orientations are important shapers and carriers of culture, then culturally based and community-based science education should work to increase identification with and ownership of science, as students see their own values and orientations embraced in curricula and associated practices. However, the changes in identification and ownership should not be confined to the students themselves but also should infect teachers, community members, and even the research team itself.

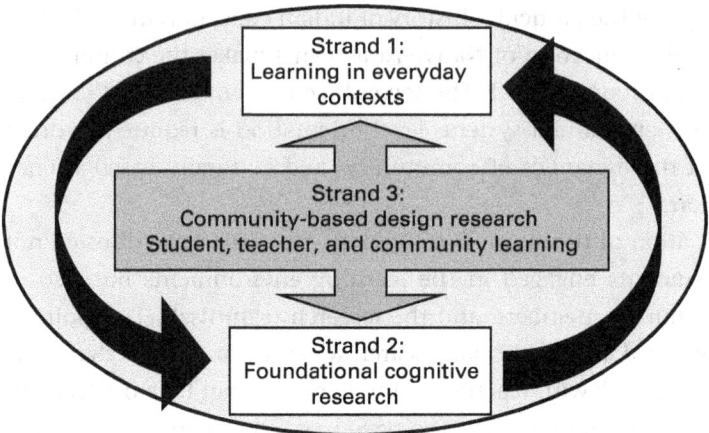

Figure 10.1
Three strands of work studying culture and cognition with respect to the natural world.

Chapter 11 reviews some of the history of Indian education that provides essential background for our study. We then go on to describe our study of culturally and community-based science education, first in the more narrow sense in which we initially conceived it and then more broadly with a more systems-level perspective. The final chapters of this monograph then focus on implications of these analyses.

11   A Brief History of Indian Education

As such, self-identification may or may not reject the "sign" Indian—or that which signifies what a "real Indian" is or looks like (often an ecology-loving, bead-wearing, feather-having, long-haired, tall dark man or woman)—and its meanings to others. Brayboy (2005, 434)

Choosing when to begin the history of Indian education necessarily has political implications. The "beginning" of Indian educational history often is assumed to coincide with "European contact." This, of course, is *not* the beginning. Indigenous people had been educating their children to become successful adult members of their communities for millennia. Perhaps the post-contact marker as the beginning can be thought of as the recognition of a significant shift in Indian education. We worry, though, that this practice has the effect of erasing or suppressing the reality of education in indigenous communities in favor of facilitating a short view and the colonial timeline. An even more serious reading of this move is that it reflects the actual stance toward Indian education—as if education only *began* with contact. We begin our narrative of Indian education with post-contact not because it is the beginning, but because it marks a significant shift in the purposes and implementation of education and because our current situations in Indian education are born of this era.

Post-contact Indian education has been an effort to assimilate Indian people into the American mainstream. Some indigenous scholars have called out these efforts as part of the ongoing intellectual genocide of indigenous peoples (Warrior 1994). More generally education imposed in a variety of forms—e.g., through the narratives or histories that are told or not told—has been a central place in which whites have sought, implicitly and often explicitly, to define Native peoples in ways that benefited whites.

The narratives and images of Indians produced by the mainstream have shifted from time to time for different (non-Indian) purposes. For some purposes of environmental education, the constructed image is that of an ecology-loving, feather-having, longhaired, tall dark man like "Iron-Eyes Cody," featured in a Keep America Beautiful television campaign in the early 1970s, with a single tear running down his cheek, whose purpose has been to remind whites of their environmental shortcomings. (Ironically enough, Iron-Eyes Cody was not an Indian but rather an Italian actor.)

For other purposes such as undermining treaty rights, whites argue that Indians should be treated just like them. For example, although the U.S. Constitution is a living document, the U.S.-Indian treaties established much later are seen as "out of date," often because they are perceived to "give" Indians "special privileges" rather than being seen as binding agreements that enabled the creation of the United States. In still other circumstances blatant racial slurs and other forms of racism (e.g., mascots) are seen as socially acceptable when targeted at Native peoples and in some minds are seen as "honoring" Native Americans. While the forms and purposes for white-produced images of Indians shift, they all fundamentally work to erode and assault, intentionally or unintentionally, indigenous sovereignty and self-determination. We'll take up some of the issues associated with these views at the end of this chapter. For now, we want to focus on schools.

Before going into details on schools and schooling, we want to provide a frank overview of a history that often carries over to the present. You may have heard that Indian children are "just not competitive," that an Indian school student would rather get a C and be like other students than get an A and excel over classmates. Some scholars ascribe this tendency to the importance of community over the individual. Although this idea may have merit, it implicitly treats schools as innocent and, if anything, as victimized by this collectivist bias.

But let's try an analogy. Change the setting to World War II and Nazi Germany and suppose that Jewish children were being sent to special schools to be educated with the explicit and implicit message that the Aryan way of life was superior in every way to that of other races. (For the record, we're not endorsing this objectifying construct of race. Race is a social construct, not a biological one, and what may seem to be obvious racial boundaries in one country are interpreted differently in other countries. In the same way there may be historical changes within a country—for example, at one point in the United States, people from southern Europe were not considered to be "white.") You can elaborate this analogy in various ways, but you can see that it's far from obvious that Jewish children in this scenario would identify with the school system or that their parents would push them to get superior grades, even if there were some benefits from doing so.

Is this a bad or inappropriate analogy? Maybe. Maybe not. Arguably, the United States has engaged in actions (e.g., the Sand Creek massacre; mass removals like the Trail of Tears or Longest Walk; or sterilization of Native women without their knowledge or consent) that resulted in genocidal outcomes for Native peoples. There are many instances of events, policies, and whites' attitudes across the United States that encouraged the killing of Native Americans. For example, in 1853 the state of California created a bounty on Indian scalps, a policy that continued up to the beginning of the twentieth century. The school system can be seen as an extension of this policy, focusing on, as some scholars put it, "colonizing" children's minds. One could argue that the United States, until the self-determination era, fluctuated between policies set on terminating Indian peoples' sovereign status, for example through termination and relocation policies which included educational opportunity as a feature of relocation, and policies set on assimilating Native Americans into the mainstream, for example through boarding schools or through mass removal of children (sometimes referred to as the "Lost Bird era"). This removal sent Native youth into predominantly white foster homes in the 1960s and 1970s based on accusations of child abuse. This so-called "abuse" was often defined by claims of cultural differences in childrearing practices or not complying with compulsory attendance laws for schooling. The Indian child welfare act (ICWA) was eventually created to prevent this form of forced assimilation. The school systems characteristically have been positioned as tools in both policy stances. In fact, much of education focused on Indians was defined by Captain Pratt's infamous statement, "Kill the Indian, save the man." The boarding school era of the twentieth century was intended to do just that by engaging in the forced removal of children from their families and communities under the guise of educating them.

A great deal of scholarly work has explored the brutalities of the boarding school era, in which Indian children died in great numbers and severe physical, emotional, psychological, and sexual abuses were inflicted on Indian children in the schools (e.g., Lomawaima 1995, 2000). Depriving children of their families, community, culture, religion, values, and language was the explicit aim of these efforts. There have been substantial changes in education policies directed at Native children, but serious problems remain. Atrocities are often seen as something from the distant past. Well, they just aren't. For many people this is the experience of their grandparents or even later during the "Lost Bird era" mentioned above.[1] Many of the Indian parents of children in K–12 education today experienced this trauma, along with associated educational, physical, and emotional abuse.

There are also enduring, significant issues that are ongoing and not confined to the past. Language is perhaps the most obvious example; most Indian children go to schools or have access to learning environments that are taught in English only. Further, most schools serving Indian children, whether they are tribally controlled schools or not, abide by national rather than Native American standards. Given that we are in the era of "tribal self-determination," there is something amiss with this. Imagine if China or any other country determined the U.S. public education system, its learning goals, assessments, language of instruction, and more. Neither the U.S. government nor the general population would be happy with it.

If this were your collective history and experience with education, would you be single-mindedly focused on having your children do well in school? Perhaps not. Remarkably, many Native nations and communities have demonstrated extraordinary resiliency and a balancing of goals such that success in mainstream education and the vitality and engagement in their own ways of knowing remain focal aspirations. Despite the historical justification for disengagement from public education, academic achievement remains a central goal for many Native peoples and a high funding priority for tribes. Education is often framed as important to improving lives and protecting and maintaining community. Although we're going to talk about science education from Native perspectives, much of what we will have to say likely holds for or is relevant to indigenous education in general.

Education from Native Perspectives

The story of science education from Native perspectives necessarily comes from a distinct frame of reference. Understanding the history of science education from Native perspectives means understanding the evolution of formalized schooling, with its changes or impositions of new forms of knowledge construction and new (and often alien) frameworks for understanding the natural world. There is much to say about this and we will continue to elaborate on these issues later. Here we provide a brief survey of the larger shifts occurring in indigenous education.

As mentioned in the beginning of this chapter, pre-contact (pre-contact with Europeans, that is) education in indigenous communities is rarely considered in the history of Indian education. Many lives' work could be made detailing the educational practices of indigenous communities from a historical perspective. For our purposes, we first point out that education in indigenous communities did not start with missionaries or the U.S government. There were sound models of education that had worked for tribes for thousands of years to ensure economic, social, and political vitality.

Although many of these models of education may not have been formalized in the way we think about them now, some were. For example the Choctaw had formal school systems for their children pre-contact (Szasz 1977). Post-contact, the federal government would look to these schools as models of success. Many people shared the role of teacher within a tribe. At the same time, there also were individuals who had proven themselves expert in certain areas or who undertook specialized training, and they would be charged with educating all children within a band or tribe (Szasz 1977).

Although there was great diversity across tribes in the practices that needed to be mastered by youth, many Nations shared the practice of using apprenticeship models for teaching. It was standard in Native communities for young people to observe tasks for long periods of time before attempting the tasks themselves. Interestingly, there are contemporary educational researchers who are exploring and documenting this learning and developmental dynamic in various settings (e.g., Rogoff 2003). Most communities had elaborate developmental pathways for their children and clear markers for, and recognition of, accomplishment.

The insertion of religious and federal education into Native education marked a dramatic change in the learning process for Native children. It sought to terminate, devalue, and delegitimize the use of indigenous knowledge in all of its forms (Noel 2002). Within communities, Native people have always had highly developed knowledge of natural phenomena as well as sophisticated technology that improved the quality of life for tribes (Deloria and Wildcat 2001). This wisdom, as we noted earlier, sometimes labeled as "traditional ecological knowledge," is continually being reevaluated, accepted, and, perhaps more importantly, sometimes recognized for its accuracy and efficacy by the "mainstream" American culture. (As an aside, this recognition in itself demands reflection.)[2] But we're getting ahead of ourselves—let's turn to post-contact "education," starting with mission schools.

From at least the seventeenth through the nineteenth century, the goal of Christian missionaries and the U.S. government was to displace indigenous knowledge and values in all forms by "educating and Christianizing the Indian." The major components of this education included speaking, reading, and writing English, with a primary focus on speaking so that Indians could participate in church services (Provenzo and McCloskey 1981). The intent of destroying indigenous knowledge has been termed intellectual genocide and is arguably one of the most destructive forces that Native communities have endured (Warrior 2000).

The establishment of the United States changed these Christianizing educational efforts in important ways. By the early 1800s, with the appearance of the Monroe Doctrine, "promoting civilization among the aborigines" became a central focus of the U.S. government. The thrust of the associated curricula included reading, writing, and arithmetic, as well as a focus on agricultural techniques and Christian religion (Sharpes 1979). The federal government also tried to change hunting and gathering communities into farming communities, conveniently forgetting that they very often had forced tribes onto land not suited for agriculture. During this time the federal government was also subsidizing mission schools to continue the Christianization process. These goals dominated education efforts until after the Civil War.

Perhaps the most well-known counterexample to federal education efforts is the story of literacy within the Cherokee Nation lead by Sequoyah (Foreman 1938; Bender 2002). The impact of written text on the unfolding of the history of the Americas is extraordinarily complex. Sequoyah, respecting the ingenuity of written systems, was convinced that written text was something that all people could master and that it could be useful to his own people. Quite remarkably, he developed an eighty-six-symbol system (now eighty-five) for the Cherokee language, and this Cherokee written language spread very rapidly in the Cherokee Nation. Within just four years the majority of Cherokees were reading and writing. In short, the

Cherokee very likely had the most effective educational system for literacy development this continent had ever seen. Literacy rates among the Cherokee were substantially higher than among the surrounding "settlers." However, the introduction of U.S.-run schools led to the demise of Cherokee literacy and academic success, replicating a pattern of failure of U.S. efforts throughout Native nations.

Following the Civil War a further shift occurred, marking a new era of Indian education. Previously the federal government had not forced Indians to attend schools, but rather assumed that Indian people would voluntarily attend them, change their beliefs and practices, and easily assimilate into the American mainstream. As land increasingly became a sought-after commodity, the government's desire to reduce allotted lands increased, as did its impatience with the lack of Indian participation in the mainstream. By the 1880s the government had decided to force education and a specific model of culture on all Indians (Stahl 1979). The creation of boarding schools became the answer to the problem of "educating the Indian."

The boarding school era was devastating to Indian communities. Many children were forcibly removed from their families and communities and transported great distances to intertribal boarding schools (Peshkin 1997). The conditions in these schools were horrific—conditions that would ultimately cause the federal government to close them down. The precise death toll of Indian children during this time is still unclear (Heart and DeBruyn 1998).

Boarding schools continued the goals of teaching reading, writing, and arithmetic, but also placed great value on Native children learning European cultural norms and practices. Boarding schools also introduced trades based on gender identity to Indian youth; for example, girls were predominantly instructed in housework and sewing (Szasz 1977). Here's an example that will make things concrete. A Menominee elder once shared stories with Medin about how the wife of the (white) School Superintendent tried to get Menominee girls to become "proper ladies." She described being brought to a tennis court where the girls were supposed to play tennis in long dresses, as was the custom in that day. The elder told Medin, "I just climbed a tree to watch and laugh." But of course the intent behind the goal of creating proper ladies was not at all funny.

The Meriam report (1928) marked another new stage in Indian education, one that is still very active. The Meriam report denounced boarding schools and called for dramatic improvement in Indian education. By 1934, legislation had been passed to petition states and territories to partner with the federal government for the education, medical care, and social welfare of Indian children (Stahl 1979). Essentially this legislation permitted Native American children to attend public schools; it is still what permits reservation children to attend public schools off reservation, as well as allowing for the creation of public schools on reservations. By the 1930s, the curriculum delivered to Indian children was theoretically not different from other children's; however, some Indian historians have argued that assimilating Indian children into mainstream American ideals was still at the heart of education.[3] This legislation also provided for penalties to parents who did not send their children to school (Sharpes 1979) and contributed to the mass removals of children in future decades.

Over the next forty years the requirement that Indian children attend school would remove more Indian children from their families than any other factor. In many ways this legislation allowed for a new form of boarding school. The U.S. Department of Child and Family Services (DCFS) removed Indian children from their homes and placed them with white families because Indian parents were not forcing their children to attend school (Peshkin 1997).

There are several generations of Indian adults alive today that spent at least several years living with white families and attending public schools away from their homelands. Eventually Indian activists and lobbyists called enough attention to these practices that a new law, the Indian Child Welfare Act of 1978, was created to stop the massive removal of Indian children from Indian communities. This law provides guidelines and appropriate procedures for DCFS that ensures that even if an Indian child is removed from his or her parents, that child will usually be placed with either relatives or other Indian people. Only in very rare instances will Indian foster children be placed in non-Indian homes (see www.nicwa.org).

Of course it is one thing to pass a law and quite another to enforce it. As we write, there continue to be stories of the widespread practice of placing Indian children in orphanages rather than placing them in foster care in Native families. (See for example Sullivan and Walters 2011.)

The American civil rights movement would mark another era in Indian education. Two major pieces of legislation started to change the educational landscape for American Indians. The first, the Indian Education Act of 1972, part of the Elementary and Secondary Schools Act, appropriated monies for remedial programs, bilingual programs, and teacher training for teachers of American Indians, and training monies for graduate and professional studies and facilities improvement. This was important because it provided funds for state-run schools to specifically serve American Indian needs.

The second piece of legislation, the Indian Self-Determination and Education Assistance Act of 1975, led to the emergence of contract schools. Contract schools involved independent Indian school boards contracting with the Bureau of Indian Affairs (BIA) to run their own schools (Noel 2002). This was the first time that there was any form of local control of the educational process. The 1994–1995 school year was the first time that there were more tribally controlled schools than BIA-controlled schools (Tippeconnic 1995). However, given the facts that (1) there are 565 federally recognized tribes and only about one hundred tribally controlled schools, and (2) almost two-thirds of the American Indian population does not live on reservations, non-Indian entities and institutions are still primarily in control of educating Indian youth.

Culture in the Classroom

Over the past thirty to forty years of Indian education, scholars and practitioners have been struggling with the best ways to think about and see culture in the classroom. Demmert and Towner (2003) wrote an extensive review about the ways in which culture has been taken up in educational research.

From their literature review Demmert and Towner propose six critical elements of culturally based education: "(1) recognition and use of indigenous languages, (2) pedagogy that stresses traditional cultural characteristics and adult-child interactions, (3) teachers' pedagogy congruent with traditional culture as well as contemporary ways of knowing and learning, (4) curriculum based on traditional culture that places children in a contemporary context, (5) strong community participation, and (6) knowledge and use of the social and political mores of the community" (Demmert and Towner 2003, 10). Other scholars have argued for a seventh proposed component: incorporation of relationship to place and values associated with land (e.g., Cajete 2000; Kawagley 1995; Ledward, Takayama, and Kahumoku 2008).

There is a growing literature examining these issues in depth and exploring the negative consequences of ignoring them. For example, Philips (1988) did remarkable work focusing on the ways in which gesture, response time, and functions and relationships to silence (among other dimensions) impact education for Native children. She demonstrated that educators' privileging of particular culture norms for gesture, response time, and verbal performance had significant effects on the educators' perceptions of children's learning, engagement, and ability, and conversely had negative consequences for children's perceptions of classrooms, learning, and engagement.

There are three case studies and contexts we will highlight that have demonstrated the power of developing culturally based education, including efforts by Native Alaskans, Native Hawaiians, and the Navajo Nation. These three have been well studied and have had a longevity that is particularly noteworthy. In each of these cases, education systems have been rebuilt from community-based perspectives and have intentionally rooted themselves in the culture and language of communities, but also simultaneously worked to engage contemporary contexts.

Perhaps the most developed model, though, has emerged from Maori communities in New Zealand. Three entire school systems have been developed and implemented at the national level: a normative system in which Western culture and content was the foundation, a language-based system in which students learned the same basic content but in a Maori medium, and a third system based in Maori language and culture (Reid 2000).

In all of these cases the development of culturally based education has been aligned with the increasing assertion of sovereignty and school improvement. These efforts are indicators that we are witnessing the emergence of a new era in indigenous education, one in which self-determination is the leading paradigm (Tippeconnic 2000).

If the history of Indian education seems like a sad history, it is. This chapter has described the history of (assimilative) education of Native children in the United States; we have difficulty seeing why anyone would expect Native children to do well in a system that is so opposed to Native values and orientations. Before moving on, however, there's one elephant in the room that bears some attention. Oddly enough the elephant is often well camouflaged and it is easy to miss. It concerns race and culture.

Tribal Critical Race Theory

In what follows we're going to rely on the scholarship of Bryan Brayboy (2005) and we're going to state the central issue and conclusion right off the bat. "Native American" is treated as a racial entity for many purposes, but Native Americans also comprise many sovereign nations (as we mentioned, there are 565 federally recognized tribes and numerous other state-recognized tribes) with treaty-based relationships with the U.S. government spelling out rights and responsibilities for all parties. Often people wittingly or unwittingly subvert the latter relationships by focusing on race. Racism is an endemic and ongoing issue, but for Native peoples it commonly gets interwoven with the denial of sovereignty in a pernicious way.

Critical race theory focuses on racism and the ways in which the constructions of race and racism are structurally embedded in our legal, political, social, and moral systems. These systems function in ways to keep racism a normative feature of American society and to support white privilege. Numerous education scholars have suggested that these issues play out in education contexts, where issues of racism and inequality are reproduced through the explicit and implicit privileging of whiteness. In addition, the reproduction of narratives of nondominant communities are often damaging at best. Tribal critical race theory (TCRT) makes a number of complementary claims and we focus on four of them (see Brayboy 2005).

TCRT asserts that:

1. Colonization is endemic to society.

2. U.S. policies toward indigenous peoples are rooted in imperialism, white supremacy, and a desire for material gain.

3. Governmental policies and educational policies toward indigenous people are intimately linked around the problematic goal of assimilation.

4. Tribal philosophies, beliefs, customs, traditions, and visions for the future are central for understanding the lived realities of indigenous peoples.

Brayboy goes on to analyze how the U.S. imperialist mindset led to policies like "Manifest Destiny," the idea that individuals have a right and moral obligation to use lands considered "vacant" because these were not being "improved." This policy was blind to Native peoples' relationships with land (e.g., the idea that land wasn't the sort of thing one could own any more than your grandmother is) and blind or opposed to Native peoples' land management practices. Even modern practices may be so entrenched that their role in preserving the status quo goes unnoticed. For example, although "white supremacy" nowadays is seen as radically racist and condemned, status quo policies such as "legacy admissions" to college[4] continue to be seen as natural and legitimate at the same time as affirmative action policies are being attacked and dismantled.

TCRT argues that Native peoples face the challenge of being both racialized and having a status as legal/political beings. Neither works to the advantage of Native Americans; the former is associated with racism and the latter, as we have noted, typically gets conceptualized as a guardian/ward relationship between the U.S. government and Native nations when trustee/beneficiary may be more appropriate. And perhaps most ironically, these two wrongs make a third wrong—obfuscating Native American legal rights under a trustee/beneficiary relationship by shifting the discourse to one of race, implicitly demolishing the legal/political dimension.

We cannot hope to do more than touch the surface of TCRT. It is multidimensional and far-reaching in its consequences—for example, when participation as a member of a legal/political entity is based on "blood quantum," a racialized marker. But we do *not* see this discussion as a diversion—the intermixing of race and culture is a complex issue for all peoples of color, but its particular historical association with the legal status of the relationship of Native nations with the U.S. government adds another layer of complexity. As we will see in the next chapter, the efforts of Native entities to reclaim sovereignty play out in the domains of education in general and science education in particular.

12   Culturally Based Science Education: Navigating Multiple Epistemologies

Today's warriors need to be armed with a PhD.
—Former Menominee game warden

There's a sense in which this chapter is out of order. We're going to present our work on culturally based science education as if the steps from theory to practice were trivial. That is, our focus will be on how we used cultural differences in epistemological orientations to develop and test science curricula. Our general idea was to present science in form and content likely to match the epistemological orientations that Native students engage the world with in their everyday lives.

Here's an analogy. Anyone who has traveled from the United States to the United Kingdom and wound up walking about London's streets has experienced some disorientation, because the attentional habits appropriate for negotiating traffic when cars drive on the right do not work so well in a situation where cars drive on the left. The same sort of disorientation may hold when science-related practices that Native children bring to school mismatch the science practices that are recognized and honored in the classroom. For example, a child who has learned that it is important to provide the context before "getting to the point" may find that when she sets out to answer a question, the teacher listens briefly but then moves on to another student who might answer more directly.

Although the goal of improving science learning is relevant to the communities within which we are working, it may not be obvious (initially) whether and how this is pertinent to or supports our overall thesis. But it is and it does. The argument goes like this: if different epistemological orientations toward and with nature affect how science gets done and if these practices are also reflected in curricula and how science gets taught, then science learning should also be affected by learner epistemological orientations. We have documented cultural differences in epistemological orientations and, on our view, they should affect science learning, precisely because science is necessarily taught from an epistemological orientation, and a culturally infused one at that. If science were just a science, without a point of view, then there would be no strong reason to think that science instruction favors a Western, middle-class value and belief system. In other words, our studies of culturally based science education help to show just how culturally embedded science practices (and science education practices) are. It also almost goes without saying that there are other factors outside of science instruction (e.g., history of relationships with schools, institutionalized racism, cultural practices not specific to science, previous learning opportunities, even nutrition) that may be associated with cultural differences in science learning and achievement.

In this chapter we will first consider some background issues, then situate culturally based science education in terms of epistemological frameworks, and describe our intervention. Our focus will be on the students and on changes associated with their participation in our programs. Note, however, that this description leaves out a great deal of relevant detail. In the next chapter we provide some of that detail and it will become clear that "community-based" is a more accurate description than "culturally based."

Background: Underrepresentation in STEM Disciplines

Student achievement in science education is a well-rehearsed problem, particularly for those groups of children who have historically been placed at risk. This problem is particularly acute with indigenous populations. As we noted much earlier, various estimates suggest that Native Americans are underrepresented by a factor of at least three or four. To make these numbers concrete, between 1998 and 2007 a total of only fourteen doctorates were awarded to Native scholars in computer science, ten in physics, five in astronomy, three in ocean sciences, and one in atmospheric sciences; in the biological sciences, 108 doctorates were awarded to Indian scholars. These numbers only represent 0.3 percent of the total number of degrees awarded (NSF 2007). The numbers represent an increase in representation from earlier decades, albeit a minimal one.

The "places" of learners and practitioners of science from nondominant groups are increasingly a focus in analyses of science learning and education in the United States. More recently this dialogue has shifted from performance to knowledge of STEM (science, technology, engineering, mathematics) content and the ability to think critically about this content. It has also included attention to learning environments such as museums and other out-of-school learning environments. Although the ability to think critically about STEM content is a necessary lens for understanding the challenges facing science and science education, by itself it is incomplete, because it focuses on the goal and not the nature of learning itself.

These same orientations treat learning as acultural. Consequently, in the context of issues of minority scholars being underrepresented in the sciences, they lend themselves to deficit orientations. At a minimum they imply that "equal opportunity" is all that is needed. Beyond that, prescriptions for addressing underrepresentation may take the form of thinly disguised (or even overt) efforts to get children and parents of color to adopt white middle-class practices and orientations (Nisbett 2009).

We believe that central to the future of science and science education is understanding, supporting, and leveraging the various ways in which diversity—of people, practices, languages, meaning, knowing, epistemologies, goals, values, and the like—is an asset (Gutiérrez, Baquedano-López, and Asato 2000; Warren et al. 2001). According to this framework, we must understand learning and development as fundamentally cultural processes (Cole 1996; Lee, Spencer, and Harpalani 2003; Rogoff 2003; Nasir and Hand 2006). Among these cultural processes are epistemological orientations.

Epistemological Orientations

Although other researchers have demonstrated that learning involves more than cognitive processes—identity and affect are intertwined—we see epistemologies as a key component of cultural processes, having impacts on engagement and identification (Nasir and Hand 2006). In this respect we concur with the spirit of the NRC report on informal science learning (Bell et al. 2009) in stressing motivation, fascination, and personal relevance as central factors in learning.

Epistemologies as Cultural Processes

Although we've been tossing around the term "epistemology" a lot, we haven't so far situated it within the broader framework of its use in educational research. In education, most research makes the assumption that the epistemologies that students bring to classrooms are inferior, or less productive, than the ones that researchers and educators are trying to implement in order to foster student learning. Some researchers have claimed that successful science education will require students to replace their personal epistemologies with an epistemology that is aligned with a Western scientific epistemology (Strike and Posner 1985; King and Kitchener 1994).

Within science education more specifically, Hammer and Elby (2003) suggest the alternative strategy of framing student epistemologies as "epistemological resources." These are developed in students' everyday lives and appropriately employed in various contexts. The resources are not part of a robust, stable, or context-independent theory or belief about knowledge and learning; rather, they vary across contexts and domains, depending upon the appropriateness of fit. Hammer and Elby nicely demonstrate how even young children are able to draw on these resources given the appropriate context.

There has been a paucity of work on student epistemologies concerned with cultural differences. Is there cultural variation in the fundamental epistemological resources different individuals bring to bear in learning? The material we have reviewed over the past several chapters appears to document differences unambiguously, but we have not so far demonstrated their relevance to science learning. Before doing so, we review some previous work on indigenous science that provides a key motivation for our research.

Indigenous Science/Science Education and Epistemology

Issues of epistemology are a rich area of scholarship for indigenous people working within a variety of disciplines (Waters 2004). A body of scholarly work has described and analyzed the plethora of ways in which ethnocentrism plays out, especially with regard to epistemology, indigenous traditions, Western European traditions, and those that have emerged from them (see Barnhardt and Kawagley 2005; Battiste 2002, 2008; Cajete 1999; Deloria 1998; Deloria and Wildcat 2001; Hermes 2000; Grande 2000; Lomawaima 2000; Meyer 2001).

There has been less work within the context of science and science education specifically, although the work that has been done is extremely important (e.g., Kawagley 1995; Cajete 1999; Pierotti 2011). Scholars such as Cajete (Cajete 2000) see Native science in terms of epistemological stances and values, not simply as part of tradition but as alive and relevant today—not confined to so-called "traditional ecological knowledge" which honors the past but also positions Native science as irrelevant to the present. Meyer (1998) nicely frames the importance of epistemology in relation to education: "Epistemology, the study of knowledge, is the starting point for any discussion of indigenous education. It is also a discussion of the priorities and need for identity. Understanding what Native peoples believe about their knowledge origins, priorities, context, and exchange teaches us more about its continuity. Knowing something, then, is a cultural experience that strengthens or fractures culture" (22).

Native children necessarily must be able to reconcile and navigate competing epistemologies and culturally inflected practices at a variety of levels. For example, their understanding of the core biological concept of "alive" must shift depending upon context. In one study we gave urban Indian middle school students a series of sixteen pictures (of animals, plants, water, sun, rocks, artifacts) and asked what a science teacher would say is alive and what an elder would say is alive. Generally, the students answered differently for each context, saying, for example, that an elder but not a science teacher would say that rocks and water are alive. Among other things, this observation reveals that Native students recognize differences in orientation, which raises issues concerning how different orientations are coordinated or negotiated by students.

Given that science instruction is seldom recognized as a set of cultural practices, many Native students may perceive a sharp divide between everyday practices and what takes place in school. The lack of recognition of science and science education as being a set of cultural practices may implicitly or explicitly teach Native students that their own orientations and practices are not recognized or appreciated in school contexts or relevant to professional science. Consequently, it may be hard for Native students (as well as others) to resist the view that science is indeed a practice peculiar to white males and that science learning consists of the "received wisdom" of the dominant culture. That's not a prescription for engagement with science.

Our culturally based science education efforts have focused on identifying and supporting (Native American) student epistemological orientations (Bang and Medin 2010). We turn now to a more detailed description of this project.

Community and Project Contexts

Earlier we gave a general description of our communities, and we supplement that a bit now.

Rural Menominee Wisconsin Community

The Menominee children in the study attend a tribal school. The majority of the teachers and staff and all of the children are Native American. Although exposing children to the Menominee language is an important focus of the tribal school, science instruction and everyday discourse is in English. Outside of school there are opportunities for activities like berry picking, harvesting milkweed, collecting ginseng, and taking forest walks. For many Menominee community members hunting and fishing are important activities, and many children are familiar with both by age twelve.

Urban Indian Population

Our learning environments are implemented in and around the American Indian Center of Chicago (AIC). The AIC is the oldest urban Indian center in the country and serves as the social and cultural center of the Chicago Indian community. We will refer to the community members from the Chicago community as the urban Indian community. Readers should note that this collapses significant cultural and historical experiences of the larger multitribal community. This could potentially have an unfortunate homogenizing effect. At the same time it would be inaccurate to suggest that there are not shared practices, values, challenges, and strengths that are fundamentally defined by being a part of the Chicago Indian community.

Now that we have tiptoed around the topic of pan-Indianness, we should bring in a more assertive perspective. Pierotti (2011, 4) says:

This supposed critique seems to be employed as a means of deflecting or distracting Indigenous people from arguing that perspectives and philosophical themes exist that are shared by almost all Indigenous people, and that these themes yield powerful insights into the functioning of the natural world that rival or even exceed those of the European philosophical and intellectual traditions in sophistication and usefulness. I have never heard it argued that there is a problem with a pan-European body of knowledge.

However, we will offer one final hedge: to the extent that this view undermines sovereignty, it too becomes problematic. Still, we do not think that trying to understand indigenous peoples as developing bodies of knowledge that can be seen in relation to one another does this.

Unpacking Epistemologies and Querying Stereotypes about Science

Developing culturally based science curricula is far from straightforward. One of the key aspects of our work has been the evolution of our understanding of what culturally based science programming means and the ways in which to design and study the programs. "Culture" and "science" are two concepts that are strongly subject to stereotyping and simplistic definitions. For example, it may be easy for some people to think of science as a body of knowledge and to imagine scientists as (white) men wearing white lab coats and using beakers and test tubes. Similarly it is easy to think of culture as a set of ideas about what people think or their customs rather than as affecting how people think.

If these stereotypes and reductionist approaches remain unchallenged, then it is natural to take some preexisting science curriculum and build in a cultural connection by "adding culture to it." Indeed this is an approach that has been widely advocated and used, but has failed to have the desired impact (Hermes 1999; Yazzie-Mintz 2007). In part, we think this is because this approach has neither addressed the core problems of culture in science and science education nor recognized the embedding of culture in everyday practices.

We think that cultural practices and their connections with Native ways of knowing must be the foundation of a community-based science curriculum. Recognizing the significance of Native epistemologies may remove some of the problems with student navigation of ethnic and academic identities that are documented in the literature (e.g., Nasir and Saxe 2003) and put students in the position of successful "border crossing" (Aikenhead 2006; Gutiérrez 2006).

It is time for us to get a little more concrete. A significant focus of our project was the creation of curricular units developed by the Chicago and Menominee community-based design teams. The design process is described in more detail in chapter 13. It included participation of a range of community members including elders, parents, teachers, youth, and other community members interested in the project. The overall process included a series of community forums and meetings over a year and half to conceptualize and contextualize the overall research project. Where relevant, we shared with the community the results of our studies reviewed in the earlier chapters of this monograph. The community members took charge of developing learning goals as well as identifying and nominating community design leaders and teachers. From these larger discussions the nominated leaders and teachers met weekly or biweekly to develop specific activities and lesson plans. These materials were shared, edited, and revised with the larger group in an iterative process. Community members also were involved in developing, refining, and collecting all sources of data.

The design principles developed in our work are aimed at reinforcing a relational epistemology for engaging with the rest of nature. At both sites a consensus developed to employ the following design characteristics: (1) using local, place-based instruction, and hands-on experiences (see Schroeder et al. 2007 for a relevant meta-analysis); (2) linking program practices with community participation and practices including community values, needs, language, and experiences (Cajete 1994, 1999); (3) seeing nature not as an externality, apart from humans, but rather seeing humans as part of nature; (4) organizing practices around the idea that everything is related and has a role to play in the universe (systems-level or ecosystems thinking); (5) considering phenomena from a seasonal/cyclical perspective; (6) presenting science from an interdisciplinary or holistic perspective and inviting the learner to view phenomena from multiple perspectives; (7) exploring and addressing the relationships and tensions between Native science and Western science (e.g., Cajete 2000); and (8) placing science in social policy and community contexts that highlight the need for participation and leadership (e.g., Aikenhead 2006). Notice that these design characteristics correspond closely with claims about relational epistemologies as well as the empirical evidence on Native American and European American differences that we have described.

Implementation

To test our hypotheses, we developed and implemented community-based summer science programs that were designed to support students' navigation among multiple ways of knowing, including their community-based orientation. The involvement of community members in the program and the explicit use of Native epistemological orientations in science-related practices serve as a strong signal that science is not just for other people (see also Heiman 1997; Di Chiro 2004.)

On a more specific level the students were often invited to take the perspective of an animal (e.g., "put on your deer ears"). The curricula included a range of content concerning plant and aquatic life through a series of hands-on experiences, guest speakers (e.g., elders and professionals working in relevant fields), and "labs" (e.g., testing pH levels of water samples). At the AIC we used the medicinal garden surrounding the building as an anchor and then branched out to various local neighborhoods to identify and experience urban ecosystems, local forest preserves, and lakefront restoration sites.

On the Menominee reservation the focus was on the forest and waters, but the program included activities like maple sugaring and visiting the Menominee water treatment plant, which maintains its own laboratory for water quality testing. Another specific element of the curriculum was the inclusion of culturally based stories that convey some knowledge about nature, primarily stories about plants and animals.

The following quotation is from a brief vignette (Bang et al. 2010, 576) that exemplifies the kind of activities that were designed and implemented. Although there are some particulars to this activity, generally our designers followed a similar structure and logic for all of the activities.

The Chicago program was based on plant ecology and organized around the big idea that everything is related. Students "recognized their relatives" by engaging in close study with one plant species that was in the medicinal garden surrounding the AIC. Students "remade a relative" by interacting with the same plant daily in a variety of ways including daily visits and offerings, growth observations, learning, plant anatomy, soil observations and testing, and plant health and relationships (for example, was there evidence of insects or other animals interested in their plant).

These practices were integrated into other activities. For example, one activity was centered around understanding buckthorn's (an invasive species) impact on local forest ecosystems. We went to a local forest preserve, accompanied by forest preserve staff (practicing scientists) where buckthorn is damaging the health of oak trees (and thus the forest canopy) and ultimately the entire health of the forest ecosystem.

Upon arriving at the forest preserve students were first introduced to the history of the preserve and Native peoples' relationships with the forest before European contact and how that relationship changed over the course of U.S.-Indian history. Through this history students were introduced to their community responsibilities to the forest and to the respectful protocol for entering into special places. They were also asked to locate their plant relative in the forest and to make a series of observations about the plants focused on their habitat, anatomy, proximity to other plants, and of the plant's state of health.

After each student located their plant we gathered together to learn about buckthorn from that plant's perspective in order to strategically clear (cut) some of the buckthorn. The idea of invasive species was reframed as distant relatives who had lost their relationship with people. We wanted our students to have continuity in orientation toward plant life even if the plants were not part of traditional Native communities. Students learned appropriate community-based protocols for cutting down these plants, safe and proper use of tools, as well as species identification strategies for various stages in a plant's life cycle. During this time we were visited by a doe and fawn walking through the preserve. The elder on our trip interpreted this as the doe and her fawn welcoming us and thanking us for the good work we were doing. Students, teachers, and other community members then cut buckthorn for a couple of hours. There was also a series of mini lessons that took place about other local plants, plant identification and plant anatomy. We were fortunate to observe several other animals during the visit including a possum and possum baby sleeping in the trunk of a tree, a snake, mice, and squirrels.

Navigating Multiple Epistemologies

In the preceding vignette one can see multiple epistemological orientations being supported. For example, the labeling of learning about plant ecology as "Remaking Relatives"[1] places the foundation of student learning in a community-based epistemology in which plants are relatives. This decision was readily associated with community epistemologies because students were in a medicinal garden that incorporated plants that tribes have used for various purposes for millennia.

Each student was asked to "visit" with his or her plant relative daily, and this included making observations of their plants through a variety of means and senses. In addition, some standard science data collection practices, such as measuring plant growth and testing soil pH levels, were included as well. As a way of making visible boundaries on appropriate types of data collection from a community-based perspective, students were asked not to collect specimens from their relative plant during this process. Teachers and students discussed the value of not collecting parts of a plant unless it was necessary. Students did harvest plants when making medicines and distributing those medicines to community members.

As the students were engaged in a new place—the forest preserve in the vignette—they were first taught about the history of the place and their ancestors' relationship to it over time. Knowing place over ancestral time is a critical component of community-based ways of knowing (Cajete 1999; Kawagley 1995). Connecting urban Indian youth to place invited them to see Chicago as Native land (albeit ceded territory), where their ancestors had been before. In addition, program designers drew students' attention to how different orientations toward land led to different uses.

Extending the frame of plant relatives and human relationships with plants to invasive species served two important functions: (1) it demonstrated to our youth (we say "our" here and elsewhere and discard the typical separation between the researchers and the researched by affirming our own identification) that community epistemologies can be expansive and not just restricted to our medicinal plants, and (2) it invited understanding of human interactions with ecosystems and the idea that people have an impact on them (or more properly, a role to play within them).

Program Outcomes

Although our goals included the children acquiring a body of knowledge concerning the natural world, we saw that as a byproduct of our overarching goals of supporting Native relational epistemologies. Our primary focus was on students' perception of science and their relation to it. Central to our design was the premise that Native students will be more engaged in school science if they see it as relevant and useful to their communities (e.g., Aikenhead 2006). Further, we hypothesized that students would take ownership and participate as expectant apprentices if they understood science as a set of practices closely associated with or used by tribes both historically and currently. Based in part on previous research (e.g., Lederman et al. 2002), we designed program interviews that explored content knowledge, conceptions of the nature of science, and associated motivations for and identification with science. We conducted pre-program and post-program interviews over two summers that included both open-ended and rating scale questions.

There were several notable findings from our pre- and post-interviews, and the results we report held for both the Menominee and the AIC implementations (again, see Bang and Medin 2010 for more details). For the rating scale questions, children showed a reliable increase in their willingness to endorse the statement, "My tribe has been doing science for a long time." This is supporting evidence for the claim that students shifted their stance toward science as something done by Native people. Note, however, that it also reflects a shift in ideas about where scientific knowledge comes from.

Pre- and post-interviews were coded for sources of science knowledge. Students showed a reliable shift from saying that they learn science in school and from books and teachers to expanding their sources of knowledge and contexts in which they learn science. Students' post-interviews included community as a context for learning science and enlisting people in the community (elders, parents, relatives, and ancestors) as sources of science knowledge.

The following is an example from our pre-interview with Sarah, a sixth-grade Choctaw student who was born and raised in Chicago.

Interviewer: How do you learn about science?

Sarah: Well, I learn science by my textbook about how different chemicals can change from liquid to solid and how earthquakes how they began how it shifts and then it cracks open and cut the earth in half and how hurricanes and twisters become and how the twisters become to the tornado how hot air and cold air blend together.

Interviewer: Who teaches you about science?

Sarah: My teacher, her name is Mrs. Smith.

Interviewer: I have a friend who says you can learn about science by watching television. Is this right?

Sarah: You can learn some stuff from television but not all you can learn from a textbook. At school they provide us with science videos to watch and it teaches us a lot about how twisters and hurricanes and liquids.

Before her involvement with the science program Sarah squarely located science and science learning as a school-based activity. There is no hint of practice-based orientation toward knowing science in her answer. Interestingly, she qualifies her answer about whether television can serve as a source of knowledge by also locating the viewing within a school context.

Contrast Sarah's initial answers with those in her post-interview.

Interviewer: How do you learn about science?

Sarah: By my elders and my mom and teachers.

Interviewer: What sorts of things do they teach you?

Sarah: They teach me about how a long time ago my ancestors, how they used to like plant and if there's weeds how they would get it out. They burn . . . when plants use[d] to take over they would burn all of them down in one spot.

In her post-interview, Sarah identifies her elders and her own mother as well as her teachers as sources of science knowledge, reflecting a shift away from viewing science solely as a school-based activity. In addition, Sarah begins to reflect a historicized view of science knowledge by including her ancestors as sources of science knowledge.

In another post-interview example for the same question, Rachel, a seventh-grade Lakota student, says: "Sometimes my school, sometimes my parents, sometimes I just discover things on my own . . . pretty much just go for a stroll. You can learn about science by just looking around and seeing what is happening. Watching ants grow or working actually watching it—that would take months, but . . ."

From our perspective the inclusion of school, home, and community life as well as their own selves as sources of science knowledge is perhaps the most empowering orientation our students could take up. This shift was mirrored in students' conceptions of the nature of scientific knowledge. When students were asked how they would explain what science is to someone who lived far away and was never exposed to it, they showed a significant shift from talking about science as facts or a body of knowledge to talking about science as a set of knowledge-making activities done in school and community by Native people.

The final result we'll mention is the change in the form of knowledge students demonstrated in pre- versus post-interviews. When asked what constitutes a forest in pre-interviews, students tended to give lists of kinds (i.e., trees, plants, animals, dirt, water). In the post-interviews students named specific organisms (e.g., milkweed, arrowroot) and typically described some relational property associated with them. For example, in a typical post-interview, a student might list a plant like poison ivy and note that deer eat it or mention that there are certain specific plants that grow by a bog because the soil and water they need is nearby. Interestingly, there was also an increase in the form of student explanations from single actors to chains of events, one marker for systems-level reasoning.

The following example is from a section of Seth's post-interview eliciting reasoning about a forest ecology scenario. He is a sixth-grade student who is Ojibwe, Navajo, and Lakota. Students were asked how they would know if there was an overpopulation of deer in a forest and what the consequences might be. In his pre-interview Seth says that he didn't think anything would really happen to a forest if there were too many deer. Post-implementation Seth says, "Well, there would be less plants because deer are herbivores and there would—and I would be seeing a lot more deer and a lot more—the other animals wouldn't be around as much because there is too many deer and there is not enough plants to feed them all." This change in the form of content knowledge reflects students' change in perception of scientific knowledge being about facts to descriptions, explanations, and narratives.

These are striking results for a three-week summer science program. They invite one to imagine how student engagement might change if a relational epistemology were at the heart of an entire science curriculum.

Summary

Although our project has aspects specific to indigenous communities, it can be seen more broadly as another demonstration of the potential efficacy of culturally based education (e.g., see Lee 1995, 2001 for other examples). The close convergence of our results on psychological distance and a relational orientation with program outcomes for identity and engagement supports our general theoretical framework.

But there's a lot more to the story. We have argued that the foundation of *culturally based design* rests on the comprehensive participation of community members, including teachers, elders, parents, community experts, researchers, and youth in all aspects of the research, from conceptions of the problems and project design and implementation to data collection, analysis, and dissemination. So far, however, we have only discussed student outcomes, and although these outcomes are encouraging, it seems odd (very unrelational) to ignore other critical aspects of a systems-level analysis. We address this limitation in the next few chapters.

13 Community-Based Science Education: Menominee Focus

We have to take control of science so that our values are reflected in forestry practices.
—Justin,[1] Menominee community member

Chapter 12 glossed over many issues and in some important ways was misleading. It was misleading not only because we only focused on student outcomes, but also because we omitted an analysis of factors that may be crucial to on-the-ground success. One way to think about this is reflected in the titles of that chapter and this one: we went into our project thinking we were going to do culturally based science education by working with our communities, but eventually we realized that we were engaged in community-based science education, with culture coming into play as a necessary corollary. We don't think this is a subtle distinction, but we recognize that it may not be obvious from the onset. Chapter 13 aims to provide more of the context for our project so that the role of community members becomes more salient.

The term "community members" may be (psychologically) distancing and therefore misleading. When we think of the "average person" we don't attribute to them any special expertise or skills. In our research communities there is considerable and highly varied science-related expertise based on both formal and informal learning. For example, on the Menominee reservation a typical community meeting will include teachers, loggers, forest ecologists, and other fishery and forestry employees. Several of these community members have master's degrees (e.g., in forestry), but more relevant is the fact that they have decades of experience working in the forest, hunting, trapping, berry picking, and the like. The meeting also would include elders, members serving on the school board, and at least two former tribal chairs. A corresponding American Indian Center (AIC) meeting would include elders, a hydrologist, an ethnobiologist, experts on the history of Indian education, community hunters and fisherfolk, singers, community artists, a former assistant dean at a local college, teachers, undergraduate students, and a student or two enrolled in master's or PhD programs.

As a consequence, community meetings might include wide-ranging discussion, drawing on personal observations, stories, teachings from elders, and a back-and-forth dialogue about forms of evidence and their support. For example, at one Menominee community meeting where the discussion centered on tribal efforts to regenerate white pine, the discussion included a historical quote from (Menominee) Chief Osh Kosh on how to manage the forest, concerns about avoiding "mono-cropping" and honoring the multiple functions of the forest (including its sacred sites), and a critique of a University of Minnesota study on white pine (by a tribal member with a master's degree in forestry from the same school) because it relied on ideal growing conditions that either would not be found in more realistic circumstances or would be too costly to achieve.[2]

An AIC meeting might include the story of how maple sugaring was discovered and developed from three or more tribal traditions as well as a discussion of how to prepare our children to deal with the possibility of being the only Indian child in the classroom. The latter might be accompanied by a story from a parent about how her young child felt conflicted between the home teachings that plants are to be respected and you should talk to them and his teacher's stern admonition, "No, no, we don't talk to plants."

Project Leaders

Over the past three to five years the project leaders have been Karen Washinawatok and Shannon Chapman on the Menominee reservation, and authors Bang at the AIC and Medin at Northwestern University. There are many, many other key players, and we apologize for not providing a corresponding amount of background information on them (though we will say more about them later on in this chapter).

We've already introduced ourselves, so let us say a bit more about Karen Washinawatok and Shannon Chapman. Washinawatok has a lifetime of experience on the Menominee reservation, having served as tribal chair, as a member of the tribal legislature, as dean of NAES College (Native American

Educational Services), Menominee campus, and she has been a research partner since 2003. She is currently director of the Menominee Language and Culture Commission and a member of the Menominee Indian School District Board. She is fluent in Menominee and has previously developed culturally based learning curricula both for middle school children and for students attending NAES College. Washinawatok likes to say that she and her husband are related to almost everyone on the Menominee reservation. For them, culture is not something to be put on display for outsiders but rather a way of life. Karen Washinawatok is a people person, and she has been crucial to building community associated with our project.

Here's an example of Washinawatok in action. A couple of summers ago an NSF program officer asked to do a site visit on the Menominee reservation. It was near the time of the annual Menominee Powwow and many people had relatives visiting and other obligations. Nonetheless, at the drop of a hat, Washinawatok organized a potluck dinner attended by approximately forty community members involved with this project in a variety of ways. The program officer was treated to a community discussion of the history of Indian education and personal narratives underlining the need for community-based science education (as well as a lively, very funny discussion of Menominee nicknames that ended with the program officer asking for and being given his own nickname in Menominee—we're going to withhold that name to protect the parties involved).

Shannon Chapman is a lifelong member of the Menominee community, has been a logging contractor, and was assistant principal at the Menominee Tribal School when she joined this project. (She is now principal.) Chapman replaced Carol Dodge, Menominee educator and elder, who took over the Menominee component of this project when Washinawatok took on the role of tribal chair. Chapman has excellent organizational skills and no one was surprised when she became principal at the tribal school in 2009. Somehow she manages to combine a no-nonsense business acumen with a wicked sense of humor (we will withhold examples to protect the parties involved).

These are the bare-bones descriptions and don't begin to capture the many hours spent together preparing grant proposals, discussing the details of collecting data, developing and implementing coding schemes, writing papers, and strengthening connections. Washinatok was instrumental in introducing Medin to Menominee elders and instructing him in the proper protocol for meeting them. Many Menominee believe that "the Menominee have been studied too much," and Chapman and Washinawatok have contributed greatly to community members coming to see that research can be an instrument for supporting sovereignty when the tribe is included as a stakeholder.

Just a Little More Background

A significant barrier to constructive engagement of science education research within Native communities is the physical, social, historical, and power distance between major research universities and tribal institutions. Although the more recent development of tribal colleges and universities is a positive, the primary mission of these tribal institutions has been education, not (educational) research. Earlier we mentioned the boarding school era and the enduring historical trauma associated with it and other assimilation efforts that have been integral to schools and schooling.

Just in case we have haven't emphasized the point enough, here's a quote from Captain Richard H. Pratt of the U.S. Army at an 1892 convention: "A great general has said that the only good Indian is a dead one, and that high sanction of his destruction has been an enormous factor in promoting Indian massacres. In a sense, I agree with the sentiment, but only in this: that all the Indian there is in the race should be dead. Kill the Indian in him, and save the man." The general in question was Philip Sheridan. It is ironic that Sheridan Road runs through the Northwestern University campus, and equally ironic that one of the founders of the university, John Evans, was implicated in the Sand Creek Massacre of 1864 when he was serving as governor of Colorado Territory.

Control over the education of indigenous children and even the parenting of indigenous children has been systematically and intentionally manipulated as a way to "solve the Indian problem." For many community members, memories of school are devastating. One of our community members is an elder who was forced to go to a Catholic school in Keshena. Keshena is part of the Menominee reservation (where the tribal offices are located), but nonetheless she shared with the community that she still gets an upset stomach every time she drives to Keshena because of those school memories, well over fifty years later. In short, it is not easy to conceptualize schools as a resource for community values. One reason the development of tribal colleges and universities is so important is that it reflects efforts of tribal entities to assert sovereignty.

Our research team recognized these challenges and intentionally proposed engaging teachers and community members in the design of learning environments. The aim was to begin to create a space where community members engaged in reclaiming the classroom for indigenous children's education (Smith 1999; Battiste 2002).

We will now give a brief overview of the design process and then turn to a detailed example of community-based design in practice and its effects within the Menominee community. Then in chapter 14 we will give a corresponding example for the AIC community design process.

The Design Process: The Early Stages

At our first community meetings we tried to stress three issues that were motivating our project. The first was the serious need for more Indian people in STEM fields, a pragmatic, relevant concern. We had done a quick survey of STEM-related positions across several Indian nations, and the majority of them were filled with non-Indian people, including at that time the CEO of the forestry operations on the Menominee reservation. This was framed as an issue of sovereignty because it meant that non-Indian people had a powerful influence on natural resource management policies in tribal communities—policies that often were a source of conflict.

The second issue we introduced was the achievement statistics for Native Americans in science, from Menominee children's performance on Wisconsin state standardized tests to underrepresentation in obtaining advanced degrees. We also described cognitive research we had been conducting with indigenous children, Menominee included, in which basic biological concepts and reasoning patterns were probed. As we noted, these studies suggest that Native children come to school with *advanced* understandings of biology and a propensity to reason ecologically. We suggested that the knowledge and skills Indian children bring to the classroom was potentially a productive intellectual asset that schools fail to mobilize or recognize.

The last piece of "evidence" that we introduced to this project was a brief survey of science classroom materials we conducted in which we looked at the way content was organized. We found that systems-level analyses in general and the coverage of ecosystems in particular were often presented in the last chapter in textbooks, with systems-level analysis essentially never used as an organizing principle. Most biology textbooks started with a micro level or what we call a model species level and then expanded (Bang, Medin, and Atran 2007). We suggested that this assumed trajectory of learning did not align with Native students' ways of knowing or experiencing the world. Thus we were proposing to design learning environments with an ecological orientation and community-based practices as the foundation, to see whether student engagement and learning were better in such environments.

The early stage of our design process consisted of a series of monthly or bimonthly meetings aimed at making sense of the goals of the project and developing a shared vision. These meetings soon evolved into specific decisions about content focus, activities, assessment, and new forms of community involvement. The majority of the work in the first year consisted of making sense of issues of science and science education from a larger cultural perspective rooted in a particular place (the Menominee reservation) and based on participant experiences. For many community members it was essential to review and relive their own, typically very unpleasant school experiences.

Here are a few examples. One community member who attended a Catholic school described the teacher's practice of forcing students who (in her eyes) had misbehaved to crawl under her desk and remain there while she taught. This community member went on to say that the teacher's attempt at humiliating him failed one time when he found he could reach up behind the desk drawer to get access to the candy that the teacher had confiscated—this was his one positive memory of school.

Another Menominee adult who attended school off the reservation told Medin the following: "On the first day of science class the teacher asked if anyone had any question about science. I raised my hand and asked, 'Why is it that some people tend to sneeze when they go outside into the sunlight?' Well, the other students laughed and the teacher laughed and I never again asked a question in science."

As part of a seminar with Menominee Tribal School educators, teachers were asked to draw a picture that captures their experience of their own schooling as children. One Menominee teacher's drawing of the school she attended as a child made it look remarkably like a jail, with only a single small window near the top of the building. She suggested that the window represented hope.

In a moment we will focus on selected segments from a design team meeting on the Menominee reservation at which elders, teachers, and community members were present. The design team was working on a forest ecology unit and discussing what they wanted the learning goals of the unit to be. Your first impression may be that the discussion has little to do with what should be in the forest unit, but we suspect that you will ultimately see it as central.

Before turning to the meeting in question, the reader must know something about the history and relationship of the Menominee Nation with their forested lands. The Menominee have managed a sustainable logging operation, including a logging mill, for more than a century (Beck 2002; Davis 2000; Grignon et al. 1998). The forest is more than a source of economic value; many Menominee have a deep sense of identity connected to the forest. The forest is a source of game, firewood, medicines, and berries, and a site for the tribe's cultural practices. Some Menominee say that if the forest were gone the Menominee would no longer exist as a people.

Some hunters express a sense of awe when they note that on the very same ground where they hunt, their ancestors hunted more than a thousand years ago. We once asked a Menominee hunter what he thinks about when he is hunting and his answer was, "I pray."

The Menominee relationship with the forest has been entangled with the history of the United States and the majority culture's domination of indigenous people. Even after the boundaries of the reservation were established in 1856, the tribe continued to struggle with outside interests and the federal government. The so-called "Pine Ring" attempted to steal Menominee timber and to gain control of and clear-cut the Menominee forest. Newman (1967) estimates that about one million board feet of timber were stolen from the reservation between 1871 and 1890.

The struggle has been protracted and multifaceted (Grignon et al. 1998). In 1871, the U.S. Secretary of the Interior agreed that the Menominee be allowed to cut and sell logs to mills outside the reservation. Under pressure from the Pine Ring, the government halted the Menominee logging operation in 1878. In 1882 a special act of the U.S. Congress allowed the tribe to cut "dead and down" timber. In 1888 the U.S. Attorney General ordered another halt to logging on grounds that the timber was government property. In 1890 another Congressional act authorized cutting and sale of timber under the supervision of government superintendents. In 1908 a bill authorized the Menominee to build their own mill and to harvest mature trees under a selective cutting system in which U.S. Forestry Service specialists would mark the trees to be cut.

In 1912, agency superintendents began a policy of clear-cutting in direct violation of the 1908 act. Selective cutting was reinstituted in 1926. In 1928 the tribe was able to elect an advisory board, and the board went to Washington to protest the mismanagement of the forest and mill. It took until 1951 for the tribe to win an $8.5 million settlement for the failure of U.S. officials to carry out provisions of the 1908 act.

At present, Menominee Tribal Enterprises (MTE) manages the logging operation in the forest, and the current CEO is Menominee. But even now cutting proscriptions are subject to approval by the Bureau of Indian Affairs (BIA).

At one point, the Menominee nation was one of the most economically successful Indian communities in North America. Despite federal oversight and mismanagement, the Menominee logging operation employed hundreds of Menominee people and generated significant revenues. But in the early 1950s the federal government began a policy known as "termination," in which the sovereign status of targeted nations was removed. The Menominee termination act was signed into law in 1954 and implemented in 1961. In effect, it was an attempt to legislate the tribe out of existence. The idea was that all Menominee would become American citizens, thereby releasing the federal government from its trustee responsibility to the Menominee tribe. The tribe would receive no more financial support from the federal government, its land would be divided up, and the reservation would become nothing more than another Wisconsin county.

The termination act immediately crippled the Menominee Nation financially, because the tribe had no tax base to generate revenue. The tribal clinic and hospital soon closed. There was a deep ripple effect on all aspects of the tribal community and, overnight, Menominee County became a pocket of poverty. The Menominee logging operation had to focus on efficiency rather than maximizing Menominee employment. These operations were managed by Menominee Enterprises Incorporated (MEI). Despite MEI's name, its board of directors included non-Indians who were often in the position of casting the deciding votes. In 1968 MEI engineered a project to combine several smaller reservation lakes to create a manmade lake, Legend Lake, and sold shoreline lots to non-Indians.

In this case, from the perspective of most Menominee, their loss of sovereignty over the lands far outweighed any economic gains, and MEI's decision triggered a storm of protest. This protest ultimately led to the restoration movement that achieved success in 1973 when President Richard Nixon signed the Menominee Restoration Act.

Today the logging operation is operated by MTE under control of the Menominee Tribal Legislature (though the constitutional relationship between these two independent entities is sometimes in dispute). MTE has won several awards for its sustainable forestry practices, yet its current cutting practices have not always been well received by the Menominee community. These practices include forty-acre clear-cuts and shelterwood cuts that look quite a bit like clear-cuts. Many Menominee people, including loggers, are disturbed by these cuts. They do not see the rationale for abandoning selective harvest, they note that the heavier equipment used for such cuts compacts the soil, they worry about the ecological consequences of these practices, and they find these cuts aesthetically displeasing. When community members object to MTE's practices they are often met with the counterargument that "our practices are based on forestry science."

This is a crude gloss of a troubled yet resilient history. We hope it will be enough to enable the reader to see the shared historical experience that may be functioning in the unfolding of the conversation. There are places in the conversation in which we will clarify terminology using square brackets and italics. We begin with Justin, a tribal leader, who has been involved with forest policy issues for over a decade.

Justin: There seems at this moment in time, or in the last decade or two, a great anxiety between science and the way traditionally the forest was managed, and how science is somewhat taking an upper hand at the um, anxiety of a lot of tribal members and loggers.

Justin begins by suggesting that the current problems are different from past problems but are not unrelated. He casts science and Menominee traditional practices and knowledge as in opposition and suggests that it is causing "a great anxiety." The expression "great anxiety" seems to index a historically infused emotion and perspective. Justin goes on to characterize the anxiety by linking it with issues of power and domination and locating it within the lives of community members (both "tribal members" and "loggers"). Note that there is no balance or integration of science and traditional management but rather science has the "upper hand." It is also important to note that at this point Justin has not located himself specifically within this relationship—he is narrating it from a relational distance.

Sarah: Are they doing things more scientifically here now, or is that some of the problem between. . . ?

Sarah, an educator and elder, takes up Justin's comparative frame and wants to analyze the difference in more specific detail. She uses the words "more scientifically," perhaps implicitly accepting the opposition of science toward tradition, if not science's superiority.

Justin: There aren't many tribal lumber mills that are a hundred years old [reference to the Menominee Mill which celebrated its centennial in 2008]. So they don't have a real record, the record that we do, and part of the practice, traditions, and oral history that contribute to how they drew up proscriptions [plans for forest management]. But, but science is definitely, I think, creating a lot of anxiety among different tribes *because the people who have the knowledge aren't the people who are making the public policy decisions* [italics added], and for some reason they can't talk, come together or talk, find a way to talk about it where they're understandable and make sense versus what they know to have had success in the past.

Justin appears to read Sarah's comment as suggesting that previous forest management was unscientific. Justin's response questions Sarah's ranking of scientific practices. He suggests that the scientists providing advice do not have the same record of sustainable forestry that Menominee do and that Menominee knowledge is older and deeper than the scientific knowledge now driving management of the forest. He positions himself as in opposition to outside science proscriptions both by the use of "they" and by questioning the validity of outside proscriptions.

The exchange also suggests that Justin has multiple frames for thinking about what power is and where it lies within this situation. Justin suggests that Menominee practices and traditions once determined the management of the forest, something that is missing from the current management plans. There is also a shift from speaking about science as a disembodied entity in his earlier comment to something people do or knowledge that people have.

Justin notes that the anxiety that is being felt on the Menominee reservation is not something unique. He expands his argument to something that is felt by many tribes and notes that the people in power are not the ones with the appropriate knowledge. Justin ends his turn by locating the problem in the dialogue between the people involved and their inability to understand one another. Implied in his comment is the idea that this misunderstanding is based on how scientists are not making sense in comparison to what people know from experience.

Sarah in her next turn notes that Justin has cast this issue on a broader level and wants to bring it back to the Menominee community by asking "what about here?" Several people speak to this question, talking about issues as broad as intellectual property rights and sharing of information to attitudes conveyed in interactions with decision makers that leave people feeling like they do not understand what is going on.

Daniel, a community member and a long-time logger of the Menominee forest, extends the idea presented by Justin by agreeing that the problem is one of communication. He focuses on the way in which content and naming practices are playing out in the anxiety-filled dynamic that Justin has identified.

Daniel: They're in charge of the forest [MTE]. They manage the forest, and sometimes you get the feeling from them that "we're managing it, leave us alone." Don't, you know, "we're doing it." And they're doing . . . science and technology, different things [inaudible], and they call it all different names. Now, in the past ten years when this has all been really coming to a head, they're calling it all different names, and most people don't understand what's going on.

Daniel pushes further on Justin's point that there is a problem of communication, focusing on the relationship between the MTE management (and associated BIA personnel) and Menominee loggers and other community members. At this point in the conversation the original binary opposition (between Western science and Menominee values) and the sociohistorical lens that have been dominating the conversation fade into the background, in favor of a focus on day-to-day forestry practices. Specifically, the conversation turns to the different kinds of cuts proscribed by MTE and their justifications. Daniel places himself within the situation, and then projects the voice of those in power. He also locates the developing tension within a similar time frame as Justin, but goes on to suggest that there is a deliberate obscuring in the naming practices by the people in charge of forest management (MTE) over the past decade and that the naming practices function as a form of domination.

Sarah: Because they don't let us know.

Sarah apparently sees herself in Daniel's comment "most people don't understand what's going on." She voices anger about the situation and speaks from a position of lack of agency.

Daniel: Yes. You know, there's, like we talked before, there's seed tree [cuts] and there's shelterwood [cuts]. They're basically the same thing. But, they're for different trees. It can look like the same cut. But they'll tell you that, "you know we didn't do a shelterwood, that's a seed tree." And, there's regeneration cuts and then there's, uh, conversion cuts, or clear-cuts. There's a whole bunch of different words that they can use, and, the average person looks at it, and doesn't understand it.

Daniel, who does understand all the cuts he has named, reiterates his point about scientific language as a tool for domination. There are a few questions that follow about what exactly the difference is between a seed tree cut and shelterwood cut and the differences in cuts that some of the community members who are present do not understand (even those who have been loggers). This turn in the conversation reinstates a historical frame, and Justin returns the conversation to the sociopolitical level and to Menominee people's general relationship with "science."

Justin: I think one of the problems with our anxiety about this is that we don't have a [inaudible] or we don't own the science or contribute to the decision-making process. So as a result, we're, we're allowing our resources to be managed and dictated by science and professionals and I think probably every one of our proscriptions are drafted by a non-Menominee at this point and that, we in the past had a sense of ownership and knowledge that contributed to that decision making and we don't now.

Justin now places himself within the situation and gives voice to the anxiety as "ours" in an inclusive sense and "we" in terms of owning the science. Justin suggests here that we, the Menominee people, do not own or contribute to "science" or the associated decision-making processes.

Justin also seems to be struggling with another perspective in this turn, one that questions the subordination in Sarah's comment. Justin suggests that we (Menominee) are allowing the situation to happen. Although the phrasing has a negative connotation, we read this as Justin's belief that potentially these issues are within the control of the community, whether or not it is currently evident. This shift in Justin's stance challenges the power frame that was previously in place. Now the power and choices, and potentially the criticism, are attributed to community, not to science. In the next turn Daniel, identifies his own relationship to the situation and seems to push on Justin's collective "we."

Daniel: A big problem with the proscriptions up here, I mean from working in the woods, being closely related or associated with people who still work in the woods, is, you know they feel so much that it's, the forest is being *experimented on*. This place is being experimented on, and, you know, they do a lot of good things, and I got problems with some of the things I've seen that, you know, that did work, that they were doing, it was right scientifically, but culturally, it just looks so bad. I mean the buffer zone [an effort to insure that there are stands of trees between roads and clear-cuts so that they are not so visible], is the best thing they did, and that's the best thing you guys did in stopping them from going up [to the road]. . . . The people have a say. They don't think so but they do have a say.

Daniel's response returns to a practice-level analysis of the forest management program. He begins to locate himself and his family as people who have been deeply connected to the woods for a long time. He narrates his identity in relation to the argument he is making at a more specific level of detail than before.

He states that the people who are deeply involved in the forest feel that the forest is being "experimented on" from a framework that does not take into consideration Menominee values. Here Daniel is struggling with the scientific rationale for the clear-cut, which he seems to approve of, but he also notes that the appearance of a clear-cut sits in opposition to cultural values.

Daniel also is struggling with Justin's all-encompassing comment that the community no longer feels a sense of ownership and does not contribute to the decision making. He says that the buffer zone decision was a good thing that the people contributed to. Daniel continues to push on a collective stance when he uses the phrase "the best thing *you* guys did." Given that Justin has been a tribal leader, it is not so much that Daniel is distancing himself from Justin (and others) as that he is affirming the role tribal leaders should play in managing the forest. In short, it seems that Justin and Daniel are wrestling with their more specific identities and roles within the community context (Justin has not been a logger or worked for MTE but Daniel has). This agreement that the people have a voice moves the conversation to specifying a goal for the forest unit.

Justin: And instilling that in our kids is I think is one of probably the most important attributes of this [forestry] unit.

The conception of the forest unit becomes one about having children understand the larger historical context and their role as tomorrow's leaders in relation to science within this context. Young people should have a sense of sovereignty and voice, in Justin's view.

Daniel: Yeah, and letting them know everything, you know, that they have, they're the ones who can be in charge.

In a separate but related meeting the next day Justin becomes more explicit about relationships between science and tradition. He states, "We have to take control of science so that our values are reflected in forestry practices." This statement both recognizes the binary opposition between Western science and Menominee values and dissolves it. He goes on to say "and we have to teach our children that they need to take ownership of science so that they can control what happens to our forest."

Summary

This discussion hardly touches at all on any content that might be associated with the forest ecology unit. But it does reveal an understanding of science and its practices as deeply embedded in values and policy decisions. The issue of "ownership of science" is the essence of community-based design and, in retrospect, it is difficult to imagine having culturally based science without it.

The discussion we have reviewed has potentially dramatic impacts on the design decisions that could be made and the priorities for learning. Justin's stance could lead to a unit that privileges understanding of history and meanings of sovereignty and its ties to ownership of science. Daniel's stance could lead to a unit that engages students in learning the scientific language and being able to evaluate and challenge practices at the day-to-day level. Both points of view enhance and deepen approaches to learning, and both incorporate the idea that knowledge of science is a responsibility. In our Menominee program implementations we noted several times when community members told children that learning about the forest was part of their responsibility as Menominee.

Although this fragment of community discourse shows a progression from a third-person to a first-person perspective and a shift from viewing science and tradition as a dichotomy to forms of knowledge that can be integrated, the story is not so simple. Thinking in terms of dichotomies— such as school and home or science and tradition—reflects well-engrained habits. It's not so simple as "add community members and stir," because we all carry stereotypes (e.g., that science is pure and objective) that need to be undermined. These themes have been constantly revisited, challenged, and reconstructed throughout our program. In the next chapter we provide a glimpse of community-based science education at the AIC.

Now we say the drum is the heartbeat of the earth, so if you want to ground children into having a connection with the earth, it's to have them listen to that drum, just listen to it, and pretty soon their heartbeat is going to fall in with that.

—Rose, AIC community member and teacher

Obviously, there are differences in the issues faced by an urban Indian community as opposed to a reservation-based tribal nation, including differences in implementing community-based and culturally based science education. For example, the Chicago community does not have to deal with the Bureau of Indian Affairs (BIA) in establishing forestry practices (though it does need to negotiate with the Forest Preserve District of Cook County with respect to practices in forest preserves). Another contrast is that urban Indian children are very unlikely to have Native teachers, and Indians are very unlikely to be represented on Chicago-area school boards.

However, issues of Native versus Western science and relational epistemologies are extremely relevant at both sites. We should also note that some of our efforts have involved AIC children and adults coming to the Menominee reservation for activities like maple sugaring, and Menominee children and adults have returned the favor for Chicago-area activities like visiting the Chicago Museum of Science and Industry or going to a forest preserve. (Incidentally, you can get much closer to deer in a Chicago-area forest preserve than on the Menominee reservation, in part because of the chronic overpopulation of deer in the former.)

There is a handwritten poster on the wall of our research meeting room at the American Indian Center listing the design principles mentioned in chapter 12. For ease of reference we repeat them here: (1) using local, place-based instruction, and hands-on experiences; (2) linking program practices with community participation and practices including community values, needs, language, and experiences; (3) not seeing nature as an externality, apart from humans, but rather seeing humans as part of nature; (4) organizing practices around the idea that everything is related and has a role to play in the universe; (5) considering phenomena from a seasonal/cyclical perspective; (6) placing science in an interdisciplinary or holistic perspective and inviting the learner to view phenomena from multiple perspectives; (7) exploring and addressing the relationships and tensions between Native science and Western science; and (8) placing science in social policy and community contexts that highlight the need for participation and leadership. These design principles serve both as a guide and as a summary of several years of our community-based design meetings. To most readers, however, these notions are fairly abstract, so in this chapter we illustrate them by drawing on selected transcriptions from a designer meeting held in January 2007, about two years into our project. The main focus of this meeting was on teaching science as a set of practices and on the relationship between Native science and Western science. These issues arose as part of the planning of activities for our summer science program, but in neither case were these topics part of an official agenda. Instead, these concerns were both spontaneous and frequently revisited at other points.

Neither author was present at this particular designer meeting (Medin was at a meeting on the Menominee reservation and Bang was in Boston and phoned in for part of the meeting). Keeping in mind our use of pseudonyms, Mary is responsible for leading the meeting, Mike, Walter, Rose, and April are community members and teachers, and Sara, Pam, Laura, and Connie are community members. Steve is an Ojibwe elder visiting from Minnesota. Megan is Megan Bang.

Mary: So I don't know how you guys want to begin. One of us can write on the board while somebody types it up.

Mike: Do we all, going through the ones that we've already kind of jumped on, we had the adopt a plant which we changed its name during a teacher's meeting but did that one and then we also had the ball, twine activity. [This is a game involving a ball of twine aimed at illustrating the interdependency of everyone.] There're some of the stuff that we already jumped in on like the big idea, we are all related, and then the big questions which we pounded that stuff out the last big meeting. We decided on those and how those fall in line with the rest of the structure.

Mike reiterates some of the design themes (3 and 4 to be specific), including the idea that learning about our surroundings entails recognizing our relatives and (re)-making relationships. We relate this mainly to underline the point that the design principles were actively used as a guide.

Rose: How do we recognize our relatives? I think we have to keep in mind also, another important point that we talked about previous to this one, is we need to express these concepts that we're putting together for the kids in Indian thought because what you see on there is we're really fishing around for the correct English words to express the Indian thought. In creating this curriculum we also have to use our Indian thought to create our own language of how we're going to express these concepts and what we want our kids to learn and understand as well as to help us to be able to become familiar with that language, so that when we do have to give a presentation or we have the opportunity to talk to people in the future about what we're doing then we'll have that language. And it will, because we, as Native people, we have that connection, that non-Indian people are searching for. They say recycling and all of these terms, whereas we say we're living in harmony and we recognize our relatives, stuff like that. But they're not to that point of recognizing any relatives. They're at the point of knowing that you have to recycle in order to help the environment and to stop the, what is that, global warming. They're not talking about helping the earth heal. They're not talking about helping our relatives to survive. Those kinds of concepts are what we talk about and the people who study about the birds and all of that, they come from a different concept also, but they still don't recognize the birds as being relatives. They look at the birds as being a very important part of the cycle of life that keeps the earth in balance. But we have that missing piece that we need to find words, how to put our thoughts down and create that language that we need in our curriculum.

In our more formal terms Rose is addressing the issue of how differing epistemological orientations affect ways of thinking about and relating to the rest of nature. Concretely, she expresses Native epistemologies as placing Indian people as part of a system and she sees non-Indians as conceptually outside the system and acting on it. And she expresses a clear value judgment about which system is superior when she talks about what is missing from the non-Indian approach. Rose also believes that language itself carries or supports different orientations and expresses frustration that using English rather than a Native American language gets in the way of that. The language issue has been a challenge in our urban, intertribal context, and the next several turns address this problem.

Laura: Why can't we use Native words?

Rose: We can do that, I have no problem with that, but we have many languages.

Sara: I don't think we're going to go into a fluent thing. No. Your basic corn, little simple words, that's so resourceful to use. It doesn't have to be this—it would be nice if we could all . . . depending on the different nations that are sitting there we could still tap into . . .

Mike: It's like putting that in the different lessons like that.

Sara: I guess that would be another part of that when you throw the ball or the yarn saying who you are, your tribal affiliation perhaps, I don't know.

Walter: On that one we thought about changing the title so it's—and we know it's a ball but we wanted to call a relationship activity or something else; the ball and twine is just what we use to . . .

Rose: And like practical and sacred for instance, try not to keep those—help the kids understand about that too, because we have our practical language and we have our sacred language, we have our practical cultural concepts and we have our sacred traditions. That's another area where we have to really keep them clear. And with that throwing of the twine I could say *ma Lakota*, which means I am Lakota. I could say *woha Lakota* which puts a sacredness meaning to it. And that combines the whole way of living in a sacred manner.

Laura: This may be really trivial, but why don't we give these activities Indian names?

Mike: I think that'd be cool. Yeah, the lesson titles and be showcasing a different tribe with each lesson so we're not—it'd be representative of who's around the table.

Rose: I was wondering when we'd get to that point. I think that's okay. If we're going to say we're all related, we would have to say *Mitakuye Oyasin*.

Walter: We could change the ball twine activity to that, I think that would be good.

Rose: (addressing Walter): How about you, your language? What do you think?

Walter: *Eewamakiki,* for we are all relatives. So a basis for a lot of what we say, we're saying, *awaesainsaw,* which is our word for birds but also culturally the same word from *eewamakiki* we're all relatives too, but we're saying about birds is they are younger relatives. Well, it's got *eewamana,* in it now which is . . . Algonquin language is changed with front vowels all the time, but we say *awae* is just a relative, but when we say we're all related basically to others we say *eeweemakiki,* so it's like plural third person [inaudible], but when we talk about birds we say *awaesainsaw,* which means our little relatives. It implies that we're all related.

Laura: I like that. Wow, that's actually a very good point, like if you're talking about in your own language and you're—say it again, I'm sorry.

Walter: You find that in most any languages that there's some kind of comparative form of that but definitely how it comes about. The animals, in particular, those family names or relationship names are all part of the generic names of what animals are.

Rose: Who is representative in our community? We could pick out some of the languages and then we can ask different people. Like we could ask Nora about Navajo or Ojibwe, you guys [crosstalk]. Who's Ojibwe here? Megan is Ojibwe . . . have her.

Laura: Yeah, but we're focusing on a tribe each day, right?

Mary: I think just as long as we keep that in mind we can name the lessons now but that's something that when we leave today that could be a homework assignment for somebody to figure out.

Why should our community care so much about language? We can see a number of compelling reasons. First, there is increasing evidence that language affects what we attend to, how we attend to it, and the meanings we make of it (e.g., Gentner and Goldin-Meadow 2003). In the preceding examples one can see a number of metaphors (or literal claims) for relationships in play that imply that humans are closely connected with the rest of nature. More broadly, we have argued that culture affects epistemological orientations and associated aspects of thinking. Assuming that this is true and that language is a carrier of culture, it follows that language should affect science-related practices. Third, language is a marker of identity and power relationships. Although we were quite limited in our ability to employ Native American language in our programming, it was striking to see how strongly our children latched on to what we did use of it.[1]

The discussion now shifts to a review and analysis of practices and activities that bear on other design principles. Again we should emphasize that the design principles often did not dictate the practices so much as the design principles emerged from an analysis of practices. For example, the perspective-taking exercises we will describe were initially spontaneous and only later on did we appreciate how they naturally fit with the Native epistemological orientation we have described in earlier chapters. Note also that the focus on observation and perspective taking as science practices conflicts with their deemphasis in most discussions of science learning.

Mary: So, if we were to think about the lesson that we're talking about, the five senses, as we wanted that introductory lesson from the first day that was what Susan had done, I'm not sure if everybody was here when we reviewed that video and the transcripts were passed out before. If we think about that then, we can start plugging it in here. Just go over it from right now . . .

Rose: Do you want me to read the one about what she [Susan] says about the cat? "Let me ask you another question. Who here has a cat? Do you know someone who has a cat? One thing about the cat that they go from one room to the other or if you let them into the house at all and let's say they want to go, if you open a door sometimes a cat will just stand there. Have you ever seen that happen? They don't, it's like what's the problem, the door is open, move it. What they are doing is what you just did. All of their senses are taking that whole world in, all at once because that cat with whole instinct not to be taken by an adult or a wolf or lose its shot at a mouse and then go hungry for the rest of the day. So they stand their eyes in their location, their ears like the deer ears start moving around, they're sniffing the air. What do I smell? Do I smell something I can run and eat or do I smell something that is going to try and eat me? They are feeling too, not like the raccoon but they are feeling with their whiskers. They will stand there—you know cats . . . they will stand there and feel which way the wind is blowing, is it raining out? And before they are ready to go out they will put that little foot out there and those whiskers. Those little hairs on their front feet will tell them if it's hot, cold, wet, dry; am I ready to go out. Okay, I'm ready to go out." That's what she said about using her senses.

Walter: So if we're going to write the whole review for that what would we say?

Mary: We can also just jump ahead to other things like how is this lesson connected to the big idea. We wrote it up on paper so we can refer back to it. Walter is the only one who can't see it. [laughter].

Walter: These are old eyes you know.

Rose: Our big idea is we are all related, right? Living in relationships, remaking our relatives, is that what it says?

Walter: So the question then is how is the five senses activity connecting to this, to the big idea of we are all related?

Rose: And another thing, just to give you food for thought too, we were talking in our teacher's meeting about some of the things that we had discussed in the last meeting and that was about how our traditions and our culture really start within us and how what this is all really boiling down to [is] learning Native science through our own eyes is really discovering all of that from within and then out. We could kind of think about that too when we're trying to make those connections between using our senses because we have that spirit within us.

Laura: I think that maybe the five senses has something to do with living— it doesn't exactly . . . well, it does fall in here but it falls in the sense that just like the cat, just like the bear, just like all the other animals, we use our senses every day in life. And I think you know, Susan putting it into those terms like we were sitting down but the kids are sitting down and they're paying attention to each sense, one at a time, you can tell that you use it every day and that you rely on it and although we are far removed from the once upon a time, [we are] using the sense of survival. I think we need to get back in contact, into that train of thought that we once upon a time did use our senses for survival, just like the cat and all of our animal relatives.

Rose: On one hand, the kids do use those senses because when they come out of their apartment they really look around, at least they should, like some do, they look around and they test to see if it's safe. Right. No one really says that out loud but actually we all do that. We look around and if it's safe that we're going to be comfortable or protected well enough from the environment, the rain, wind all that sort of thing.

Laura: And step outside see if it's cold, you go back inside and put your jacket on.

April: One of the questions that moving from big idea to big question, one of the questions that it relates to that we talked about I think two or three meetings ago, is how do things look from an ant's or a bird's or a cat's perspective? So, moving but not having just that people-dominated perspective, so I think for the senses lessons that is one of the things, it could be one of the big questions.

Here April is explicit about how the senses activity and perspective taking map on to broader goals. She implies that one of the benefits of thinking like an ant is that you may escape from just thinking like a person. That in turn may help children to engage in systems-level thinking. The next few turns build on April's observation and then link it to other science-related practices.

Rose: I know Megan thought that was really important that we look at a perspective from that point of view and maybe look at an activity where the kids could see themselves from the point of view from a bird or from a rabbit or from a muskrat.

Mary: That makes me think for that first day if we do the senses activity they could be immediately followed by the point of view of an animal or activity.

Walter: Yeah, students will be able to relate to an animal or insect's perspective of the environment or something like that.

April: The other thing the senses make me think about is recognizing our relatives. I keep remember Steve telling stories about going out to get medicine, how he uses his senses to do that.

Connie: Because when the kids go to, if they went to Waters [elementary school] they have to use all their senses to recognize the trees and not just pass up the bushes on the way back. They're actually recognizing them, aware of [them] and asking questions about them . . .

Walter: Our senses give us the goal to recognize who we are and then what everything else is and how it relates to us as our relatives. Without those senses we can't recognize it, so I think one of the goals of the activities will be that our students be able to use their senses in determining their relatives or how it all relates. Does that make sense?

April: To me that would be a goal because if you say you're going to use your senses in trying [to see], how we are all related, that's also like about an inquiry perspective, but I think part of the project is to help students engage in inquiry. So, in different cultures use their senses in different ways in determining how we're related.

Walter: So observation is key to doing science.

Note the emphasis on engagement as the key to learning. Science and science learning are not seen as activities that require people to distance themselves in some acultural and ahistorical manner. The AIC community members are implicitly following the thesis of Native scholar Yazzie Burkhart (2004, 25). "In American Indian philosophy, we must *maintain* our connectedness, we must maintain our relations, and never abandon them in search of understanding, but rather find understanding *through* them."

In her next turn Rose makes this idea explicit:

Rose: In science they use all those different test tubes and chemicals and all of that to identify and all of that quantify—all of that. So here we are, we're using our natural senses to do that. It is Native science and we could say that, we can demonstrate that. You don't have a test tube here but what would you do. We could pose a question to them like that. We don't have something to separate, quantify and qualify, but just by looking at the color of this or just looking at the [inaudible]. We were starting to look at the whole Gompers [a local park and forest preserve] through this winter and we're able to see the change.

To an outsider, an exercise focused on using the five senses may sound like some meditation technique with no obvious link to science learning. To the two authors, however, it is remarkable that the community members and teachers pay far more attention to science as a process than science as a body of knowledge. They see observation and perspective taking as fundamental to the central idea of learning about the world as making relatives and establishing relationships. For Rose, these observational skills and relational orientation are the essence of Native science. In the following segment the discussion of observation transitions into the relationship between Native and Western science.

Mary: In the back on the second page of the back page is our navigating world. That is the connection between Western science and Native science, what we've been talking about right now that could be plugged into right there. So observation . . .

Walter: Observation is a scientific skill.

Laura: I was just going to say, you can't say that Western science doesn't use observations.

Walter: No. That's the exact point we're saying, yeah.

Laura: It's the bridging point.

Rose: [Crosstalk] The bridging point is the knowledge that's been passed down about being related to all of life, being able to even learn how to structure our families in our everyday activities by using natural time.

Walter: Because the fundamental of our observation or at least from the perspective that I think we're taking is that our observation allows us to enhance our relationships with things we are observing not just to do science.

April: I would rephrase it. I wouldn't say [crosstalk] but not to manipulate that because I think we have to be careful about how we use the word "science."

The notion of a "bridging point" is more than a little ambiguous. Walter and Laura are talking about relations between Native science and Western science with observation being a common skill. Rose, however, is more concerned with bridging traditional and contemporary knowledge and practices. She may be implying a loss and a need to restore Native peoples' relation with all of life and indirectly suggesting that Western science is missing that relation. Walter's next turn serves to reinforce the latter suggestion, but April then questions the use of the term "Western science." In her next turn Rose sidesteps the issue of definitions in favor of focusing on the narrative of restoring relations.

Rose: Yeah, that's another thing we were talking about at the meeting before is developing the language that we're going to use in our curriculum and being very sure about that so we can learn it together so that in the future [we] would not only use it on the children and in our lessons but in presentations, in talking to others. I think we're going to really come up with a really Indian way, unique way of being able to do this. So it'll be cool. It'd be fun.

Do you know what too, we were thinking about, we have, I wanted to get started on a tangent up there. We wanted to include music somehow and Walter, he knows songs and he's got a hand drum. Now we say the drum is the heartbeat of the earth so if you want to ground children into having a connection with the earth it's to have them really listen to that drum, just really listen to it and pretty soon their heartbeat is just going to fall in with that. And we have a natural connection anyhow to the earth with our heart.

Rose has put her finger on perhaps the most central issue in STEM education. Science is often seen as a boring, esoteric activity that only a person who is slightly strange (the proverbial "mad scientist") would want to pursue. In contrast to that stereotype Rose perceives science learning within the frame of being connected to the earth. The next few turns reinforce that orientation, indicating just how central community members consider it.

Rose: So we have to help the kids to reevaluate their use of the senses so that they will feel the importance of that because a lot of times you don't just stop and say, oh, I'm using my senses.

Laura: Using my senses I think of haunted houses. You never use your senses as well as you do when you're in a haunted house, because it's dark and you're listening for things and you hear the chainsaw and you know what I'm saying. And you smell things, you smell smoke and you're like what's around the next corner [crosstalk] and you're all antsy. That to me is using your senses. For me, I think that is a prime example of the people we used to be. The way humans, the human spirit, the human nature, the way we are far removed from society and BMWs, that's who we are.

Rose: And too, another thing that helps you with that is ceremonies. When you go up on a hill and you have nobody there with you and everybody, you can't just look over and there somebody's there to, you can talk to or that will help you if it starts to rain all of a sudden or whatever. You're just there by yourself and when you're sitting there or standing there you just feel rooted to the earth, because there's nothing there but you and everything else. You and all of natural life and you're completely disconnected from the neighborhood, the lights and the cars and all of that sort of thing; the electricity, all of those things are just gone and then all of a sudden there you are and you just see that you're a part of all this. And somehow, somewhere, if we could get to that point for the kids to have them each get to that point of making that connection.

Laura: I think that was a major point of the spirit quests in the sense the boys out by themselves, it was to get in touch with their human nature and not only that to build relationships with their relatives.

Rose: But I like what you said about less mindful about putting a negative tone on it, it's just saying that we are mindful.

Laura: Yes, that doesn't mean to say that we, I don't think we've lost the ability by any means, it's just that it's not, we're not always conscious of it.

Rose: So when we're doing this we have to help them to be conscious of that and to have value for it. To let them know that their ability to experience the world is very, very important, to be able to get that across to them.

The conversation next shifts to details of programming activities, then returns to the observational and perspective-taking exercises as part of a Native relational epistemology. The discussion of values indirectly contrasts with Western science when the focus turns to how to show respect toward relatives.

April: So one of the big questions we talked about is how to think the former muskrat's, ant's, bird's point of view and I think another question might be what are our responsibilities to our relatives, because if you're actually going out and touching, tasting, picking things, you might want to have a conversation about how you do that or be responsible about certain things.

Rose: We kind of discussed that too at our teacher's meeting and what is a respectful way to be in terms of how you're seeing, observation [crosstalk] interact to touch. And, remember, we talked about that, about whether or not we should do a lesson on that and we said no, because we talked about giving thanks and using tobacco to give to the plants to put down and show respect to the plants. And we decided that we were going to, what did we decide?

Mike: We decided that the issue was whether or not we'd want to make a lesson about offering tobacco to plants, like whether or not that would be appropriate and then we thought it may be more appropriate to make a lesson about giving thanks rather than being, like well, okay this is—especially if it's going to be taught by non-Natives as well as Native people. You don't want to make a lesson on how to pray but you want to depict this in how to be respectful, how to do this in a respectful way, so it's about giving thanks rather than—that was one of the side conversations that we got into, definitely.

Rose: And in giving thanks then we would recognize the relatives, we would recognize them and we would be appropriate in putting tobacco down and we could talk about that and doing that so we could show the kids how to respect their relationships in doing that instead of just talking about praying, because it would be appropriate to just make a whole thing on.

The issue of showing respect turned out to be a recurring theme, in part because of our intertribal context and the differing practices concerning how to do so. Note the contrast with most science education that, though it may take up issues of the ethical treatment of animals, largely treats plants and animals as objects, not as relatives.

The designer group next transitions to a discussion of program evaluation, the use of embedded assessment, and candidates for "success indicators." They briefly consider how open-ended questions might be useful, but then turn to more practical matters such as the problems with using all the senses when interacting with plants like stinging nettle and poison ivy and what sort of instructional sequence makes the most sense. Then a discussion of clan animals leads the designers back to Native versus Western science.

April: Because clans are known for things that they can do and what they're named for. I can't think of another word except for "personify," it kind of personifies that, and so for kids to understand their relationship between animals and senses and animals and plants and why . . .

Rose: Well, we can work that in there. We can say that our people observed a deer and saw and taught our people how to hunt better.

April: But it's almost like, is "taxonomy" the right word, it becomes like the cultural way of labeling things like you would do in Western science.

Rose: So, we would have to look at clans, we could look at the Iroquois, the Ojibwa for clans and look at it and see why they were named that.

Mike: I think that that's a perfect, a good point to put in our navigating worlds [between Native and Western science] because we have to make these kinds of teaching points when we start doing the sequence, so we have notes for teachers.

April: I guess, what I'm trying to get at is that science is about inquiry and about setting up systems, and so showing students how even what we think about in just being cultural is really science. In some ways clans are scientific because it's about a system or about creating [crosstalk] a way of understanding things and that's what science is.

April explicitly rejects the idea that science is acultural. Her definition of science in terms of setting up systems puts clan systems into the realm of science. The other designers do not necessarily agree, but their responses suggest that they are intrigued by April's suggestion.

Laura: And how would you demonstrate that, how would you show that?

Walter: We can leave that as sort of one of those open-ended questions. If you pose it the right way and talk about clans and animals and their strengths and characteristics and just leave that to the kids, but then they go to the associations, because when we look at all the different cultures we're talking our background, we're talking about some of them have those clans and some don't, some of those clans represent different things for some clans, but I think it's a great way for the kids just to let them do that through inquiry; basic[ally] what characteristics of those animals that represent the bear clan. But what does the bear represent and those kinds of things.

April: And it also goes back to the responsibility question. I guess for me what I'm trying to get at is that what I hope kids take away is that science is just a particular way of observing the world, of describing the world, of figuring out how things interact, instead of thinking about science has to be a test tube so that they can find science throughout the things that they see every day.

Megan: Right, and that science isn't facts. Science is a way. I think that's our biggest difference.

Rose: That it is evolving all the time.

Rose's comment illustrates one of the frustrations with analyzing transcripts. As it turns out, her comment is not elaborated on or developed by Rose or any other group member. Is Rose expressing the idea that science is dynamic rather than a body of knowledge? Does she have in mind Native science and is she implicitly rejecting a view of traditional knowledge that may freeze it in the past? Does Rose think of science as "a way" in the same sense that Megan has just expressed? Laura's next comment may reflect a similar frustration.

Laura: I think just in our conversations we keep misusing the term "science." If we're going to say what science is, we need to say it and stick to it, because I think I've heard like five different times we've used it and used science not to mean science the way April said it, but to mean science as in Western science, as in test tubes, as in whatever preconceived ideas that we all have. Not only are we confusing ourselves, I think we're confusing the kids. I'm confusing myself and everybody in this room is confusing me, so I think we should put some definition in stone and leave it there. Find a new term. If we're going to say Western science, find a new term for it and not just say science.

Rose: What would you suggest?

Laura: I don't know. Let's use an Indian word.

April: If we say that science is a way of looking at the world and observing things, of being systematic about things, then what's overlaid on top of that is culture. So, Native Americans have a particular way of looking at things, of observing things, of documenting things. Western society has a particular way of doing that.

Laura: Exactly, and it should be distinguished. I propose that we give a Native name to Native science, to Native learning and use it. I think it would be beneficial for at least me, because I think we keep misusing the terms that we've set for science. Am I the only person that feels that way? Probably. [laughter]

April: The term "science" but I don't know that I feel comfortable giving it another name, because I think Native science is just as valid as Western science.

April is explicitly rejecting the idea that science or "modern" science equals Western science and again offers a definition of science linked to practices. She sees "science" as a superordinate term with Western science and Native science as two valid, equal-status examples of it. When Laura suggests a special term for "Native science" April rejects it on grounds that Western science should not be privileged. The discussion continues.

Laura: I'm not saying it's not.

April: I know you're not, but I think that by giving it another name we put it on a different level, higher, lower, same, whatever you want, wherever you want to place it. I think we need to communicate to kids that it's science and we all have different ways of doing science, but it's all science.

Laura: Right, but whenever you use the term "science" to mean test tubes, but then you use the term "science" to mean Native science or to mean strictly learning and one person is using it one way, another person is using it in a different way, it's not making sense, it's not translating.

Pam: I think that throughout the lessons and as time goes on if we're reiterating [to students] . . . "you know what we just did is an example of doing science or a different way of coming to know something." Then there'll be that question, oh that discussion or conversation about Native science and science as inquiry. I think the harder thing is the cultural piece to speak to that maybe.

Rose: And I think we're still figuring out our own language and how we're going to teach and how we're going to get these concepts across. And I think you're right in a way that we need to distinguish words.

Laura: Or at least we need to convince ourselves what the definition is, so that when we speak we're completely understanding the things we are saying.

Megan: I think the observation that everyone uses science to mean different things at different points in conversations is really, really important, but I also think that's how kids experience discourse about science; even in their science classrooms at school, outside of classroom, so I wonder if there's a way for us to not necessarily try to come down on solid definitions, but intentionally get kids to reflect on well, what do you mean by science when you're talking about it at this point? And well, what does someone mean by science here rather than like—personally, I don't think we're going to come up with a clear, specific definition, because I think that is nearly impossible, I think people have been trying to do it for decades and no one actually has been successful in giving a clear message about what science is in a classroom. So rather than setting ourselves up to have a definition that we then use not in that way all the time, maybe we just think about getting kids to reflect and be clear. "Wait, this isn't making sense to me, because I'm understanding science in this way," just what you just did, I guess, is make them stop and wait and figure out all of the different ways we mean science.

Rose: The intention, hopefully with the first and the second lesson would help them do that. The third lesson they'll be getting a clearer picture and then the fourth day they'll wrap it up. So the fourth lesson would be the overall point of view, so that we can get them to understand the whole process of Native science that they have experienced it during that time of learning.

April: I guess my question is, thinking about that, do we want to have the first lesson be specific about asking those questions not with coming to any type of answer, but is setting the stage for exploring the questions.

Rose: I guess the exploring what your senses [say] is a way to do that.

April: We're not explicit and I think, there's implicit lessons, but I think at some point we need to be explicit with kids about we know there's this world that you're navigating and let's have a conversation about it.

Megan: Maybe that's one of the big questions and they pose a big question about what science means. What is science?

Mike: When we were talking about science we had a few things that we— some main points, so science is about interaction, science is about observation, it's about recording, so what if we looked at a few of these main points and maybe made some more. Saying, okay, the senses activity, this is one part of science that you're doing right now. You're observing, using your senses. That's science, so you're doing science right now. I don't know if that's explicit enough, but just bringing science full circle to what they're doing right now so that senses activity it's not, the kids might be thinking, "That's not science. Why is this science?" We're not doing what we usually do, but we say no, we're observing and that's science. You observe in the classroom but you can observe using your senses. You don't need a microscope or you don't need these different components.

April: Or even to have a question about why is science so important to Native community, why do we need Native scientists? Because I think an end goal of the project is that we'll encourage more Native students to go into science, but I think you have to have a conversation with them. Like hopefully, the senses activity will excite them about science but I think it's more than just getting them excited.

We leave the conversation here. If it's not already obvious, the definition of "science" and how to navigate between Native and Western science have been recurring themes in our curriculum development and implementation. On the one hand, sometimes it seems as if no issue ever gets fully resolved. On the other hand, what we find remarkable is what does not get presupposed and discussed in this designer meeting: science as content, as a set of laws or body of knowledge. Here's our speculation on why this might have been: It is natural to think of culture as a way of life, and if that way of life embodies a relational epistemology, then culture includes a way of relating to the rest of nature.

From this perspective, Native science is not distinct from culture but rather embodies a relational epistemology. Consequently, it may not be a large step to see science in general in terms of how people relate to the rest of nature, as a set of processes and practices—as a verb, rather than a noun. Classroom science teachers may be more exposed to the idea of science as a body of knowledge.[2] Our designers and teachers did not ignore content knowledge, but science as a verb was always in the foreground.

As we reflect on these meetings and the discussions, their contrast with most discussions of and writings about science teaching and learning is striking. To be sure, there are discussions of science as inquiry and discussions of teaching science by engaging students in science practices, but, with very rare exceptions, science itself is treated as something that is beyond the bounds of scrutiny and as something that has transcended culture, if not humanity itself.

Chapters 12 and 13 show that in our communities science is seen as reflecting cultural ways of knowing and values, both explicitly and implicitly. On this view one cannot begin to do effective science learning and teaching without a close scrutiny of the cultural practices and values that are embedded in science instruction and practices.

Summary

We hope this kind of "in the trenches" examination of discourse associated with our community meetings gives the flavor of our planning process. What are not represented in our two examples are the changes that have taken place across time. We are actively engaged in analyzing longitudinal changes in teacher and community meetings, looking for shifts in discourse about the nature of science and signs of changing identification with the project in the form of pronoun shifts (e.g., from "you" to "we" and from "it" to "our").

Maybe it's time for another confession. Perhaps because of his Western science background, Medin's focus on this project often has been on experiments (and pre- versus post-program results) on Native children's conceptions of nature and of science. He saw recording designer and teacher meetings as, at best, of secondary interest. Bang, in contrast, was drawing on a systems-level perspective from the start, and for her it would have been very strange to ignore what was going on in these meetings. It is only a slight exaggeration to say that Medin was doing Western science and Bang was doing Native science.

If you are having trouble seeing how chapters 12 and 13 are relevant to demonstrating that culturally based science education is important, you too may be looking through a Western science lens. It might be interesting for you to go back and read chapter 12 after being informed by chapters 13 and 14. In the next chapter we present a broader look at program outcomes and take another look at our research partnerships.

15 Partnership in Community: Some Consequences

The concepts of creativity: chaos, participation, and metaphoric thinking, lend themselves specifically to the way that Native peoples envision the process of science. They also form a conceptual bridge between Native and Western science, although Native science refers to them differently through particular cultural representations in story, art and ways of community. (Cajete 2004, 47–48)

There is a clear link between knowledge—in the form of experience—and power. Power through an Indigenous lens is an expression of sovereignty—defined as self-determination, self-identification, and self-education. In this way sovereignty is community based. (Brayboy 2005, 435)

The foundation of community-based design rests on the participation of community members, including teachers, elders, parents, community experts, researchers, and youth, in *all* aspects of the research. This includes an ongoing conception of the problems and focus, project design and implementation, data collection and analysis, as well as dissemination and working through implications. The design process for learning environments provides opportunities for professional learning both for the teachers and designers, in our case community members.

Chapter 12 described some of our student results from pre– versus post–summer program interviews with children. But if we are thinking of culturally and community-based research at a systems level, then we might expect to see changes not only in students but also in community members, teachers, and the research team itself. Chapters 13 and 14 provide a glimpse into community-level change. To us the mere fact of sustained involvement of community members in the project is quite remarkable. Many participating community members did not have children in school, and certainly their own experience with schools would not encourage any lingering participation or nostalgia. But ownership can be powerful, and that includes our research personnel (who are also community members).

To give one recent example, Bang encouraged members of the American Indian Center (AIC) research team to work with her to develop a symposium for the 2011 meetings of the American Educational Research Association (AERA). The symposium they designed was accepted by AERA organizers. Three of the people in the symposium were neither graduate students nor faculty, but they developed insights from their participation in the project that were considered worthy of sharing with the educational research community (by peer review).

Perhaps our most surprising results are the shifts in goals for the people working on our project. From the start of this project, two Native scholars have completed their PhDs (one in anthropology, one in learning science) and another community member and teacher decided to enroll in a PhD program (at Northwestern University) and will complete her dissertation in 2013. And just to brag on her a bit, she applied for and was offered the very competitive AERA minority dissertation fellowship.

Another teacher/designer on our project and one research assistant at the AIC have completed their master's degrees. Three other teachers are working on their undergraduate or master's degrees—some in education, others in science fields. A research assistant who had dropped out of high school completed her GED and is now enrolled in college, and another research assistant is completing law school. Two Menominee research assistants who had completed associate of arts degrees years ago have returned to college to complete their bachelor's degrees. One of these two was awarded a National Science Foundation (NSF) REU (Research Experience for Undergraduates) summer research fellowship to conduct research on ecosystems at St. Olaf's College. A significant number of our research assistants attending college have indicated an interest in applying to graduate school after finishing their bachelor's degrees. Virtually all team members say that working on the project was a key factor in their decision making.

Although, at most, our project might receive partial credit, there have been a number of important changes at the Menominee Tribal School. Scores on the science section of standardized tests have improved. Menominee girls have been participating in summer science camps at the University of Wisconsin, Stout, for the past several years. The school now has a chapter of AISES (the American Indian Science and Engineering Society) and has sent teams to the national science fair (conference) in recent years. Finally, the school has an active science club.

These dramatic changes indicate a deep motivation in the Native community to take ownership of schools and ownership of research and to assert tribal sovereignty. A key factor in this change was the creation of a community of scholars. At the AIC there is what is called the "NSF room," where research team members and teachers work at their desks and hold meetings. The informal chatter can be very wide-ranging and the camaraderie is unmistakable. Clearly our project changed the college experience for a number of members of the community. It has also transformed the typical assumption that Native students need to leave community to attend institutions of higher education if their tribe does not have a tribal college that they can attend. (And many tribal colleges are only able to offer associate of arts degrees, so if a Native scholar wants a bachelor's degree or to do graduate work he or she will have to face being away from community.) These college students now have the expectation that rigorous research can be done directly in community, and they know that they can navigate between multiple contexts.

On the Menominee reservation, research assistants (RAs) worked at the Menominee Language and Culture Commission headquarters, a renovated house on the shores of LaMotte Lake.[1] Wireless Internet is available, as it is at the AIC site. One of our Menominee RAs, Brett Reiter, created a project website (http://menomineescienceproject.webs.com).

Another advantage of combining college education with work experience as a research assistant is that often students take courses for which various aspects of the research project can inform and shape academic reports to satisfy course requirements. This, in turn, gives the students confidence that they have a deep understanding of the project itself.

In many research labs research assistants and team members have specific roles and rarely have the opportunity to see the broader picture in a research project (e.g., most of the undergraduate research assistants in Medin's lab at Northwestern do not even attend lab meetings). To counter this trend, one of our major project goals is to build research capacity and infrastructure. Consequently, the RAs are expected to learn to use statistical and other software packages Microsoft Excel, IBM SPSS Statistics, and NVivo (NVivo is software for coding qualitative data) and to participate in the development of coding systems and the design of studies. Awkward as it sounds, all major design and analysis decisions are developed and approved at all three major sites (AIC, Menominee reservation, Northwestern), and our work is better because of it. All data are shared across sites. (A joint checking account may be of little value to one party if the other party retains control of the checkbook.)

Discussion

Our work has been motivated by the hypothesis that there is a dissonance between Native American cultural ways of understanding nature and the cultural ways of knowing honored in school science, and that this difference in orientations is at the heart of Native student (and often community) disengagement and underachievement. One central feature of the discord students experience is the lack of connections across the multiple contexts in which students learn science. This equally holds for grade school students who must integrate the cultural understanding that rocks are alive with a science curriculum that says they are not, and for Native college students who might dare to suggest that Native science is different from Western science but not inferior to it. Such ideas are not likely to be welcomed if the teacher thinks there is just one objective, value-neutral science.

We continue to work with and refine our understandings of what these various forms of discord and occasional harmony mean and how they might be effectively addressed as students integrate diverse perspectives. Our goal always is to develop effective teaching and learning strategies that build on the variety of resources for learning Native children bring to the classroom. An ongoing challenge is to build stable infrastructure to support Native sovereignty through research in institutions that must manage many competing goals with very limited resources. No one has ever said that they "invested heavily in supporting Native science and it failed."

The last three chapters have explored the process that enabled design teams to construct effective, culturally based learning environments. We believe that these environments have evolved to a place where there is a constructive integration of Native and Western scientific knowledge and history that is focused on contemporary problems and issues.

One of the most salient shifts in our community design teams concerns the ways in which culture is being conceptualized. Initially, many community members tended to think of culture as something of an "add-on," to be put into our lesson plans after the science part was worked out. This might take the form of "historical connections."

Now, almost all community members see culture as foundational with science built around that base. One of our teachers vividly illustrated this point with a picture she drew representing her conception of science education. She drew a large turtle (she is a member of the turtle clan) and added microscopes, test tubes, and the like on its back—very cogent indeed.

A Second Look at Partnerships and Innovations

Our ongoing project is distinctive in two ways. First, it involves a coalition of community members and teachers developing science curricula and implementing and refining curricula based on reflective practice. Second, it represents the cooperative efforts of reservation-based tribal institutions, an urban tribal institution, and a major research university. On numerous grounds this is an unlikely combination, so the synergies growing out of this project will merit careful attention. We believe that our partnerships can serve as a model for other partnerships among research universities and tribal institutions and other institutions that traditionally have served underrepresented groups. But we also should caution that it is easier said than done. There is no history of niche construction that might make this task any easier—indeed there is an entrenched environment around universities that often may be hostile toward Native scholars.

The four of us most involved in our community-based science education efforts (Bang, Washinawatok, Chapman, Medin) did not just run into each other at a coffee shop or a conference and decide to collaborate. To the extent that we have been successful, we must acknowledge coincidence, challenges, and convictions. Medin was able to start work on the Menominee reservation only because an early contact at the College of the Menominee Nation encouraged him to visit elders and gave him some important introductions, including to Washinawatok. Recall also that Bang had contacts on the Menominee reservation through friendships, AIC programming, and the fact that NAES had both Chicago and Menominee campuses. Washinawatok was dean at the Menominee campus of NAES, and our first partnership involved the two NAES campuses and Northwestern. Chapman's uncle, who was on the board of directors of NAES, became involved in our project, and saw the potential synergy that Chapman would bring to it. So connections are important and chance connections are to be appreciated.

This project has not been without its challenges. Reflective of the under-funding of Native programs nationally, NAES ran into a financial crisis and the two campuses went their separate ways—the Menominee campus was ready to close, but then was rescued at the last moment to become a campus of East–West University based in Chicago. East–West University is not a tribal entity and this long-distance relationship has not always been smooth. In Chicago our project focus shifted from NAES, Chicago, to the American Indian Center. And none of the institutions mentioned in this paragraph had a previous indirect cost agreement with the federal govern-ment or was very experienced with managing federal grants. So there have been challenges, to say the least.

But convictions thrive and grow on challenges. Throughout the insti-tutional upheavals there has been a strong group of elders and other com-munity members that have offered perspective, prayers, and passion for our project. So chance factors are important but we think they have unfolded, to quote a Chicago-dwelling Menominee elder, "in a good way."

Summary

If learning sciences research is to have significant impact in the world on the engagement and achievement of students from nondominant back-grounds, researchers must begin to recognize the real-world dynamics of community-based research. Often research projects are organized and planned around university-based models and expectations that may not be sensitive to historical factors, power relationships, community contexts, and divergent conceptions of science and science education.

Hindsight reveals that we have made many false steps. Nonetheless, our project has not only had positive results in the areas of our initial focus, but also it has had ripple effects that we did not foresee. Ultimately, self-deter-mination through community engagement with and ownership of science and science education may be the most important outcome.

It's time to return to the broader picture, sum up our arguments and evidence, and draw out potential lessons from them. That will form the substance of the final chapter.

16 Summary, Conclusions, and Implications

As all knowledge originates in a people's culture, its roots lie in cosmology, that contextual foundation for philosophy, a grand guiding story, by nature speculative, in that it tries to explain the universe, its origin, characteristics, and essential nature. Any attempt to explain the story of the cosmos is also metaphysical as the method of research always stems from a cultural orientation, a paradigm of thinking that has a history in some particular tradition. Therefore, there can be no such thing as a fully objective story of the universe. (Cajete 2004, 46)

Our overarching goal has been to do something of a systems-level analysis of relationships within and among: (1) science and its practices, (2) cultures and correlated epistemologies, and (3) science education, in order to analyze how they combine to motivate a new role for science and scientist diversity, especially as manifest in the practices of science and science education. There are several distinct components to our argument.

First, science is a cooperative enterprise drawing on many different approaches and strategies. Each science-related activity embodies perspectives and values: ideally, widely shared values, but values nonetheless. Some scholars suggest that science is self-correcting and that the sociology of science ultimately roots out bias, at least when competing biases cancel each other out. We think this view is too optimistic and ignores historical relationships between science and power. Especially when the pool of scientists is narrow with respect to relevant dimensions (e.g., gender, socioeconomic status) and shares (otherwise unwanted) biases, entrenchment may dominate and self-correction may prove to be elusive. This is the first, most straightforward argument for diversity and power sharing within the framework of the goal of producing the best science. So one important role for diversity is to undermine research biases that may be correlated with gender, race, socioeconomic status, and culture.

Second, reality—or the way things are in the world—underconstrains the research that gets done. The weak form of this argument assumes the unity of science, but points to the fact that what we can potentially know about nature is unlimited. Assuming that scientist diversity encourages research diversity, then we have the second argument for diversity in science.

The stronger but more plausible claim (in our opinion) is that science is pluralistic. By this we mean that alternative theoretical perspectives (or "styles"; see Hacking 1992) within the same domain or scope of inquiry may each yield useful insights, depending on the questions of interest and the goals and values in play. We illustrated this point with examples drawn from psychodiagnostic categories, but we could equally have done so for other systems such as taxonomic categories (again, see Hull 1988 and Lloyd 2010 for more detailed accounts and analyses). Assuming that scientist diversity is correlated with diversity in methods and theoretical orientations, we have a compelling reason to believe that scientist diversity makes for better science.

These are very abstract claims about the benefits of diversity that need to be backed up by data. We discussed feminist science in general, as well as specific examples where women have provided a distinct theoretical perspective, such as Carol Gilligan's critique of Lawrence Kohlberg's stages of moral development (Gilligan 1982). We also reviewed, for purposes of illustration, differences between Japanese and U.S. primatological research, differences that reflect (cultural) assumptions about how research should be conducted. These and related observations show that diversity and culture are reflected not just in content or interests but also in approaches and practices.

The heart of our thesis has been a detailed analysis of cultural differences between Native Americans and European Americans in approaches and practices in the domain of biological cognition and understandings of nature. We have described our comparative studies in terms of epistemological orientations and have documented cognitive consequences of these different orientations. For our Native American participants, nature is more psychologically close than for our corresponding European American participants. This difference in psychological distance affects attention to context, perspective taking, and propensity to attend to ecological relations. When coupled with a Native American relational orientation more broadly ("living in relationships"), it is associated with cultural differences in systems-level thinking and explanatory frameworks such as the abstract expectation that "everything has a role to play." In terms of our thesis, the argument is not only that culture is associated with differences in science-related practices and orientations in principle, but also that our studies demonstrate the reality of cultural differences in epistemologies.

It should be clear that we are very far from being the first scholars to describe or make claims about Native American relational epistemology. Gregory Cajete once remarked to us that "I have developed the theory, but you have provided the data." He may have been speaking about the community-based design aspect of our project, and of course the term "data" is interpreted variously in different social science subdisciplines. We are happy with the more modest claim that we have added to the data on relational epistemologies with a variety of measures and have added to the discourse about the associated implications for science and science education.

On the one hand, it should be clear that we are not making claims either about capacity or domain generality. Systems-level reasoning may well emerge in a wide range of learning environments, and European Americans certainly are capable of adopting this orientation (it may well be their preferred orientation in some contexts other than the ones we have examined). Instead our argument is that in the domain of reasoning about the natural world our Native American participants show a greater propensity to display markers of relational and systems-level reasoning. And even this claim may need qualification—we have not shown that these differences extend beyond whole-organism (macro-)biology, because we have examined neither micro- nor molecular biology.

On the other hand, the very question about domain specificity arises most naturally out of a Western science perspective. Several members of our research team, ourselves included, have questioned the separation and separability of natural and social sciences. Raymond Pierotti (2011, 199) remarks that, "in contrast to the Western tradition, in Indigenous thought there are no clearly defined boundaries between philosophy, law, biology, geology, anthropology, and geography."

The next step in our argument was to show that epistemological orientations are also an intrinsic component of science education. The principles that evolved from our community-based planning process for implementing culturally based science education again reflected Native epistemological (relational) orientations and associated assumptions about practices (e.g., inviting the children to engage in perspectives from the point of view of other species, adopting multiple perspectives focusing on observation, learning about our plant and animal relatives, etc.).

Our pre- versus post-program interview data indicate that consequences include: increased identification with science, seeing science as a set of practices rather than simply a body of facts, and understanding resources outside of formal schooling as ways of learning about science. We also reviewed evidence that community members, research assistants, and teachers came to take ownership of science and to see the potential for their values to be reflected in it. In terms of our overall thesis, we have shown that cultural differences in epistemological orientation make a difference when implemented into science education.

In summary, there are a variety of forms of evidence indicating that how science gets done reflects who's doing it. The same holds for science education. Of course it is also true that the motions that go into stirring cookie dough depend on who's doing it, and it may even be the case that how you stir depends on your culture. Our arguments and results should not be confused with this trivial truth. The case of science is more far-reaching and significant. It's not as simple as everyone doing the same thing but not in the same way—instead, it's a matter of different things being done in different ways for different purposes that are constrained by different values. So it would be more accurate to say that the very nature of the science that gets done reflects who's doing it. To be sure, all instances of empirical science are constrained by how nature is, the need to have replicable results, and so on, but within that constraint there is a plenty of room for values and epistemological frameworks to exert their influence.

Values in Science

There is increasing evidence that at least funding agencies are coming out of the closet with respect to values in science. The National Science Foundation has been considering revisions to their Merit Review Principles and Criteria. Table 16.1 summarizes the candidate criteria posted by the National Science Board on June 14, 2011.

Note that the national goals include national security and increased economic competitiveness. You the reader may agree with many or perhaps all of these goals, but you may disagree with some or think that equally

Table 16.1

Merit Review Principles and Criteria

The identification and description of the merit review criteria are firmly grounded in the following principles:

1. All NSF projects should be of the highest intellectual merit with the potential to advance the frontiers of knowledge.

2. Collectively, NSF projects should help to advance a broad set of important national goals, including:

• Increased economic competitiveness of the United States.

• Development of a globally competitive STEM workforce.

• Increased participation of women, persons with disabilities, and underrepresented minorities in STEM.

• Increased partnerships between academia and industry.

• Improved pre-K–12 STEM education and teacher development.

• Improved undergraduate STEM education.

• Increased public scientific literacy and public engagement with science and technology.

• Increased national security.

• Enhanced infrastructure for research and education, including facilities, instrumentation, networks and partnerships.

3. Broader impacts may be achieved through the research itself, through activities that are directly related to specific research projects, or through activities that are supported by the project but ancillary to the research. All are valuable approaches for advancing important national goals.

4. Ongoing application of these criteria should be subject to appropriate assessment developed using reasonable metrics over a period of time.

Intellectual merit of the proposed activity

The goal of this review criterion is to assess the degree to which the proposed activities will advance the frontiers of knowledge. Elements to consider in the review are:

1. What role does the proposed activity play in advancing knowledge and understanding within its own field or across different fields?

2. To what extent does the proposed activity suggest and explore creative, original, or potentially transformative concepts?

3. How well conceived and organized is the proposed activity?

4. How well qualified is the individual or team to conduct the proposed research?

5. Is there sufficient access to resources?

Broader impacts of the proposed activity

The purpose of this review criterion is to ensure the consideration of how the proposed project advances a national goal(s). Elements to consider in the review are:

1. Which national goal (or goals) is (or are) addressed in this proposal? Has the PI presented a compelling description of how the project or the PI will advance that goal(s)?

2. Is there a well-reasoned plan for the proposed activities, including, if appropriate, department-level or institutional engagement?

3. Is the rationale for choosing the approach well justified? Have any innovations been incorporated?

4. How well qualified is the individual, team, or institution to carry out the proposed broader impacts activities?

5. Are there adequate resources available to the PI or institution to carry out the proposed activities?

---

important goals are missing from the list. What is clear is that this set of goals expresses values and that values are on the table for discussion.

In a letter to *Science*, Robert Frodeman and J. Britt Holbrook (Frodeman and Holbrook 2011) note that "the list focuses on economics and national security but excludes protecting the environment and addressing other social problems. Aside from the consequences of neglecting these areas, this new focus may undermine the attractiveness of STEM disciplines to more idealistic students who are interested in meeting human needs rather than fostering economic competitiveness." They suggest that these changes move too far in the direction of accountability at the cost of creativity and autonomy.

To us it doesn't seem that the argument expressed by Frodeman and Holbrook is internally consistent. To be sure, the relevance to human needs of some fundamental question about the way things are in nature often may be difficult to foresee. Nonetheless, meeting human needs is a form of accountability, so idealistic students might be more attracted to STEM disciplines if things like protecting the environment and meeting human needs were part of the merit criteria, even acknowledging that payoffs are rarely immediate.

This story is still developing (see Mervis 2011) and the criticism of these guidelines continues to be bimodal. Some express concerns that the focus on diversity and improving public scientific literacy is being watered down by the additional goals, but others suggest that these broader criteria undermine basic research itself. The task force charged with developing these standards is responding to these critiques with a revision still under way as we write this. We take the open discussion of these values to be a good thing, far superior to acting as if science reflects no values.

Value in Diversity

Just one more thing about these standards. You probably noticed that the review criteria included increased participation of women, persons with disabilities, and underrepresented minorities in STEM. Alan Leshner, chief executive officer of the American Association for the Advancement of Science and executive publisher of the journal *Science*, wrote an editorial that appeared in the *Chronicle of Higher Education* (Leshner 2011), titled "We Need to Reward Those Who Nurture a Diversity of Ideas in Science." In it

Leshner suggested, "The publish-or-perish journey to tenure needs to be recalibrated if we really want faculty members to pursue and nurture the diversity of innovative scientific ideas from all students, particularly among underrepresented groups." He goes on to say, "Increasing the diversity of the scientific human-resource pool will inevitably enhance the diversity of scientific ideas. By definition, innovation requires the ability to think in new and transformative ways. Many of the best new ideas come from new participants in science and engineering enterprises, from those who have been less influenced by traditional scientific paradigms, thinking, and theories than those who have always been a part of the established scientific community."

Finally, he suggests, "the diversity dialogue too often focuses solely on a concern for equity, or just increasing numbers, and overlooks the central role that novel and creative ideas play in the scientific enterprise. Creativity is found everywhere and needs to be nurtured. New people and collaborations bring new ideas and approaches that are critical to scientific progress, even if at times they challenge long-established paradigms."

Obviously, we quote Alan Leshner so extensively because we agree with his central argument. We hope you won't think we are too churlish when we tell you that we wrote Leshner, praising his editorial, but also suggesting that it should have appeared in the most prestigious journal of science, *Science*.

We read the article on line, and that means we could also read comments on Leshner's article. The feedback was full of skepticism. A typical response was something like the following: "What is the evidence for the claim that more diversity will improve science? Claiming it doesn't make it so." Our monograph has reviewed some of the evidence supporting Leshner's claim, and our analysis of Native American and European American epistemological frameworks for conceptualizing nature provides some theoretical and empirical underpinnings for the argument.

Conclusions and Implications

We believe that the implications of diverse scholars engaged in diverse scientific practices also are potentially profound. As we noted earlier, Scott Page (2007) has shown that diversity can be more important than ability in seeking solutions to problems such as formal mathematical proofs. So we need diverse scholars in science, not only because it's the fair thing to do, but also, as Leshner suggests, because science is the better for it.

Here we need to add a hedge. Our argument holds whenever scientist diversity is correlated with a diversity of values and practices. If a field of scientific endeavor sees only one true way of doing things and one true set of values, it may succeed in suppressing alternative views and approaches. And in a field dominated by white, middle-class (male) scientists, the words "one true" may be infused with cultural assumptions and values that its practitioners perceive as acultural and objective. Under these conditions scientist diversity will not help, and minority scholars are likely to be turned off by the fact that they do not see their cultural orientations reflected in the science.

Exactly the same thing can be said about science education. We believe that one important factor in the underrepresentation of minorities in the sciences is that science education may recognize and value practices that white, middle-class scholars bring to the classroom, while ignoring or even overtly discouraging the science-related practices that other cultural groups bring to the classroom (again see Bell et al. 2009).

If this analysis is even part of the story, then underrepresentation in science will never be remedied by better schools, betters curricula, better teachers, and all other betters that leave science itself as pure and beyond examination. We are arguing that science learning and participation in science are relational in nature.

Consider an analogy with people and shoes. Imagine that science is a shoe and that, over time, the consensus developed among researchers in a position of power is that the shoe should be a size ten and only a size ten. On this view, if you want to be a scientist you have to figure out how to make a size ten a good fit for you. If you happen to be a size six or a size fourteen, you face major challenges in making this work. In the same way a (Western) science unwilling to examine its (culturally infused) values, assumptions, and privileged epistemological frameworks may work to exclude minorities and unnecessarily homogenize science and science education, because it's not the case that one size fits all.

From the perspective of indigenous scholars and scientists there may be some encouraging signs. After decades of work, in 2007 the UN Declaration on the Rights of Indigenous Peoples was passed (United Nations 2007). The resolution was approved by 144 nations; though it was one of the last nations to sign on, the United States finally endorsed it in December 2011. We call your attention to Article 31, Section 1: "Indigenous peoples have the right to maintain, control, protect and develop their cultural heritage, traditional knowledge and traditional cultural expressions, as well as the manifestations of their sciences, technologies and cultures, including human and genetic resources, seeds, medicines, knowledge of the properties of fauna and flora, oral traditions, literatures, designs, sports and traditional games and visual and performing arts."

If the United States is going to honor this declaration, it will need to respect indigenous manifestations of science and allow indigenous people to maintain, develop, protect, and control them. More broadly we need to create and support sciences for all.

How do we do that? What if scholars saw science careers as the opportunity to express their deepest values and as an effective way to "give back" to their communities? What if they saw their own culture and background as providing a distinct perspective for contributing to knowledge? And what if our science infrastructure supported diverse perspectives and science teachers treasured them? It seems to us that these are all ways of opening up science, extending an invitation for all to explore it. Our future as a species may well hinge on our ability to resist narrow conceptions of science and ways of knowing.

Notes

Introduction

1. Postmodernism in anthropology largely has been driven by nondiverse mainstream intellectuals.

2. We'll also see something of a double standard for the role of science in technology and policy issues—scientists like to take credit for the positive developments but treat any problems or unfavorable outcomes as outside the purview of basic science.

3. Note that the term "outreach," however well intended, positions minorities as outsiders. If a museum has an "outreach program," you can be sure that those being targeted are not in a position of power. One of our (Native American) Menominee friends employed by a museum is commonly asked to sit on museum advisory boards, but she says they have yet to take her advice on anything.

Chapter 3

1. In psychology we call variables like culture and gender, "quasi-independent variables," though perhaps it would be more accurate to call them "pseudo-independent."

2. fMRI provides a measure of blood flow as an index of brain activity.

3. For further discussion of perspectives on science and epistemologies see Howe (2009) and commentaries by Johnson, Tillman, and Bredo (2009).

Chapter 4

1. Stewart (2010, 46–47) underlines the significance of this sort of feedback process: "science extends its discursive reach much further, beyond technology and the applied sciences, beyond the social sciences, to influence many basic notions about intelligence, language, cognition, humanity and others that are treated as 'the truth' by society."

2. The Buckwalter and Stich paper should not be taken as the last word on this topic. Their article has received a lot of attention and at least one scholar, Jennifer Nagel, has questioned the replicability of their results. Time (and further studies) will tell.

Chapter 5

1. This description draws heavily on a review paper by Sachdeva et al. (2011).

2. Interestingly, this indirectness may have its parallels in disciplinary differences within Western scientists. One of our anthropologist friends once said to us something like (we may not recall the precise wording), "You psychologists just come right out and make a claim. In anthropology we are far more indirect; we may hint about it, approach it from a variety of perspectives, even dance around it, but we would never state things so directly. We hope that the reader will draw the relevant conclusion."

Another anecdote: Medin served on a committee to identify and address barriers to international collaborations in psychological research. The committee representation was itself international. One of the primary complaints of a famous researcher from Australia was that American scientific journals will only publish papers written in what to her eyes was a very peculiar (and we add, culturally specific) style.

Chapter 6

1. Although this comment is not directed to cultural research per se, we can't help noting that psychologists are very good at identifying scenarios where some phenomenon is clearly evident. Problems may arise when we forget how good we are at this and then assume that the phenomenon of interest is highly generalizable to other settings, samples, and materials. For example, researchers have identified a "sunk cost" effect (e.g., Larrick, Morgan, and Nisbett 1990; Tversky and Kahneman 1981) where people violate cost-benefit reasoning principles. The idea of a sunk cost is that once you have invested resources (time, money, etc.) into some activity, those resources are gone and you should only consider future costs and benefits in deciding what to do. For example, if you have paid $60 to see a play and it's obvious from the start that the play is awful, you should leave. The "investment" of $60 is not a reason to stay if you're not enjoying the play and would rather be doing something else.

Although sunk cost bias can be readily demonstrated, its generalizability should not be taken for granted. Upon reflection you might be able to think of situations where people cut off their investment prematurely (Heath 1995). For example, someone may pay $400 for ten tennis lessons but drop out after three lessons "because they aren't playing well and not enjoying themselves." Based on the sunk cost principle we should commend them, but it may be that had they continued the lessons they might have developed much more skill and greater enjoyment. In brief, sunk costs may be just one side of the coin where the challenge for people is to "know when to hold them and know when to fold them."

2. All the same one should bear in mind that it is only an assumption. Le Guen (2011) studied use of absolute (e.g., to the north) versus relative (e.g., on the left) spatial referencing systems among Yukatek Maya in Mexico. Previous work has assumed that the differences in linguistic reference terms mediated language effects, but Le Guen noted that children use an absolute system well before they acquire the Yukatek language reference system. Further studies showed that gesture was the critical factor—the Yukatek Maya use an absolute reference system in gesture. So in this case it's not a matter of language and thought but rather gesture and thought.

3. Note, however, the potential susceptibility to the home-field disadvantage. If one knows a lot about when Americans are impolite, then employing the same measures in Japan will doubtless show that Japanese people are more polite. It is always possible that two cultures do not differ overall in some abstract variable like politeness or impoliteness but rather simply differ in the kinds of social contexts where politeness is appropriate. For example, one stereotype of Native American children based on observations in school is that they are not competitive and don't want to stand out from their group. If you share this stereotype, you'll likely abandon it if you go to a powwow and watch some of the dance competitions.

Chapter 7

1. If you are an academic you may be very familiar with this situation. Someone invites you to contribute a chapter to an edited volume fifteen months before it is due, and you agree because you have something to say and perhaps you will have a graduate student coauthor who will get a much-needed publication. You hardly think about it until three months before it is due, when a reminder comes. At that point it seems like agreeing was a bad idea and that you have too many other obligations to consider. Editors also know that potential authors will turn them down if they are only given three months' notice, despite the fact that almost no one starts writing until a month or two before the draft is due. So wise editors ask at least a year in advance, when authors will not think about practicalities.

2. We need to add our usual home-field-disadvantage warning here. The body of research on dispositional bias in the United States equips researchers with lots of study materials in which (U.S.) participants ignore situational factors. We can't be 100 percent sure that, had the study of dispositional biases originated in the East, there wouldn't be good materials one could use that might not work as well in the United States. Since we have no data on this we'll offer an anecdote. If you look at "personal ads" seeking potential marriage partners in Korea, you'll see that the descriptions typically include blood type. Why? Because it is commonly assumed

that blood type determines character and personality. Medin was curious about this and called a Korean professor who is one of his former students and asked, "Why do Koreans advertise their blood type in personal ads? Is it like astrological signs in the U.S. that people mainly pay attention to for entertainment?" She responded that on the contrary, Koreans seriously believe that blood type determines personality. When Medin asked her why Koreans have this idea she replied, "Because it's true!" He pushed her a little by asking whether she knew her husband's blood type before they married, to which the answer was "of course!" She then added that she could also tell what my blood type was just from knowing me. (Without straying into TMI territory, for the record, she was correct.)

3. The Menominee traditionally have made milkweed soup. One of our Menominee friends told us his favorite meal is venison and milkweed soup. But before you go out to make it, bear in mind that the milkweed has to be at the right stage of growth or it can be somewhat toxic.

Chapter 8

1. In related work with Menominee and European expert fishers (Medin et al. 2002) we verified nominations of expertise by also asking for nominations of fishers who had as much experience and interest in fishing but not as much expertise. We then found that the nominated experts displayed more knowledge and familiarity with a broader range of local species of fish than the group nominated as having less expertise. It appears that in both communities fisherfolk have a good sense of who might know the most about fish and fishing.

2. If you're not familiar with statistics, the problem is that with zero variability, correlations that are designed to explain variability will necessarily be zero.

3. The one exception to this finding came from an ecologist who refused to sort tree names into categories on grounds that we had mixed native and nonnative species. He also railed against cultivars like the Thompson Seedless Green Ash because they are "unnatural." (Don't ask us what he had to say about manicured lawns that are maintained by lots of artificial fertilizers).

4. Of course this prediction would be bogus if we selected thirty-four pairs where there was no group difference before. In fact, however, the subset of thirty-four had yielded strong group differences in our first probe of ecological relations.

Chapter 9

1. Medin remembers chatting with a psychoanalyst friend who described psycho-therapy as always involving the same (Freudian) drama, with the only variation being that different people take on the key roles. This starts to get at the idea of

attention being focused on relationships rather than individuals, because these roles are defined by relationships in the same way that relational concepts such as "grandfather" are.

2. We do not mean to imply any single interpretation of the Menominee origin story. Some Menominees claim that the origin story is just about the development of clans and "bear" just means bear clan. Others treat the story as a rich metaphor and still others may see it as quite literally true.

3. We struggle with this terminology that carries with it an implicit understanding of nature with which we disagree. All children have equal exposure to nature unless some have found a way to travel on a different plane of existence. "Intimate contact" is a goofy term designed to capture aspects like psychological distance, salience of biological kinds in one's daily life, and diversity of experience, but none of these will hold up to closer scrutiny. As an example, Winkler-Rhoades et al. (2010) asked urban and rural children and adults (including rural Menominee children and adults) to name all the animals they could think of. Notably, urban participants tended to name exotic animals (mainly African mammals); most notably, rarely were urban, native animals (e.g., squirrel, rabbit) mentioned. Arguably, urban participants see squirrels much more often than rural participants, but rural participants were more likely to mention them.

Chapter 10

1. We say not absolutely essential because even if there were one true account of nature, the race to discover it would almost surely be faster to the extent that a diversity of strategies and perspectives were brought to bear on it (Page 2007).

2. Again we hasten to note that most of our evidence is drawn from a single tribe, the Menominee tribe of Wisconsin. Where we also have collected data from urban Indian samples in Chicago, we have generally found similar outcomes. But speculations about pan-Indian cultural orientations are just that—speculations.

3. Here we need to add a variety of hedges that alone could constitute a chapter. Our claims are about culture, not race or ethnicity, and as such they will no doubt show a great deal of within-culture variability, not to mention issues of multiculturalism or different forms of selection bias that may operate at different stages of academic training. Our claim is only that these moderating factors do not eradicate cultural contributions to science practices in general, nor do they eradicate cultural differences in epistemological orientations.

Chapter 11

1. For the story of Lost Bird, see Flood (1995).

2. There are indigenous scholars who do not see any reason for optimism, and they argue that indigenous knowledge or practices should be left in communities and not brought into schools or other arenas. Though we agree in part, we also think that there are legal developments and institutional changes (e.g., the U.N. Declaration of the Rights of Indigenous Peoples [2007] or increasing sovereignty in our own schools) that invite a more nuanced navigation through this space. Further, we resist the idea that Native children will just have to continue to survive an alien education system rather than having one that reflects their values.

3. Medin remembers one particular article in the *Weekly Reader* that was commonly circulated in Iowa grade schools when he was a boy. The article was about Indians and the question in focus was whether they should be assimilated or allowed to be different. No one, and certainly not the *Weekly Reader*, questioned the premise that this was a decision to be made by non-Indians.

4. This is the policy adopted by many universities such that if your parent attended the school, you are all but certain of being admitted yourself. Where a school has a history of excluding minorities, as almost all do, this policy perpetuates exclusion and arguably it implicitly perpetuates the associated white supremacy that accompanies it.

Chapter 12

1. "Remaking" may seem to imply a narrative of loss, but our use derived from a Lakota term, which can variously be translated as "making" or "remaking."

Chapter 13

1. Here and elsewhere we use pseudonyms. We have also lightly edited some of the quotes for smoothness of flow and to respect the wishes of tribal members.

2. See Shiva (1993) for an analysis of the costly consequences of focusing on maximizing production under ideal conditions versus success under practical conditions.

Chapter 14

1. Both at the AIC and on the Menominee reservation there are a variety of language preservation/restoration efforts in play. At the AIC there have been weekly "language table nights" where community members go to a table corresponding to the Native language they wish to speak. Typically there will be six to eight different tables. On the Menominee reservation one innovative program developed by Karen Washinawatok is a language table for elders, many of whom were not exposed to Menominee language when they were children, because their parents did not want

them to be punished for speaking Menominee in school. For many elders, learning even a little Menominee triggers poignant childhood memories (both good and bad), and the language table plays an important function.

2. In the Menominee component of our project initially we did see a somewhat greater focus on science as knowledge and culture as an add-on. This may reflect the greater involvement of classroom teachers, but fairly quickly the notion of science as practices became dominant.

Chapter 15

1. Recently they have moved into the space that used to house the Menominee County library. We see fewer eagles, but now the space is nice.

References

Aikenhead, G. S. 2006. *Science education for everyday life: Evidence-based practice*. New York: Teachers College Press.

Allen, N. J. 1995. "Voices from the bridge": Kickapoo Indian students and science education: A worldview comparison. Paper presented at the Annual Meeting of the National Association for Research in Science Teaching, San Francisco.

Anderson, E. N. 1996. *Ecologies of the heart: Emotion, belief, and the environment*. New York: Oxford University Press.

Anggoro, F. K. 2006. Naming and acquisition of folkbiologic knowledge: Mapping, scope ambiguity, and consequences for induction. Ph.D. diss., Northwestern University.

Anggoro, F. K., Medin, D. L., and Waxman, S. R. 2010. Language and experience influence children's biological induction. *Journal of Cognition and Culture*, *10*, 171–187.

Anggoro, F. K., Waxman, S. R., and Medin, D. L. 2008. Naming practices and the acquisition of key biological concepts: Evidence from English and Indonesian. *Psychological Science*, *19*(4), 314–319.

Arnett, J. J. 2009. The neglected 95%, a challenge to psychology's philosophy of science. *American Psychologist*, *64*(6), 571–574. doi:10.1037/a0016723.

Asquith, P. J. 1996. Japanese science and Western hegemonies: Primatology and the limits set to questions. In L. Nader (Ed.), *Naked science: Anthropological inquiry into boundaries, power, and knowledge*, 239–258. London: Routledge.

Atran, S. 1993. *Cognitive foundations of natural history: Towards an anthropology of science*. Cambridge: Cambridge University Press.

Atran, S., and Medin, D. 2008. *The native mind and the cultural construction of nature*. Cambridge, MA: MIT Press.

Atran, S., Medin, D., Lynch, E., Vapnarsky, V., Ucan, E., and Sousa, P. 2001. Folkbiology doesn't come from folkpsychology: Evidence from Yukatek Maya in cross-cultural perspective. *Journal of Cognition and Culture, 1*(1), 3–42.

Atran, S., Medin, D. L., and Ross, N. O. 2005. The cultural mind: Environmental decision making and cultural modeling within and across populations. *Psychological Review, 112*(4), 744–776.

Atran, S., Medin, D., Ross, N., Lynch, E., Vapnarsky, V., Ek, E. U., et al. 2002. Folkecology, cultural epidemiology, and the spirit of the commons: A garden experiment in the Maya lowlands, 1991–2001. *Current Anthropology, 43*(3), 421–450.

Axelrod, R. M. 1987. The evolution of strategies in the iterated prisoner's dilemma. In L. Davis (Ed.), *Genetic algorithms and simulated annealing*, 32–41. Los Altos, CA: Morgan Kaufman Publishers.

Axelrod, R. M. 1997. *The complexity of cooperation: Agent-based models of competition and collaboration*. Princeton: Princeton University Press.

Bacon, F. [1620] 1960. *The New Organon and Related Writings*. Indianapolis: Bobbs-Merrill.

Bailenson, J. N., Shum, M. S., Atran, S., Medin, D. L., and Coley, J. D. 2002. A bird's eye view: Biological categorization and reasoning within and across cultures. *Cognition, 84*(1), 1–53.

Baillargeon, R. 1994. Physical reasoning in young infants: Seeking explanations for impossible events. *British Journal of Developmental Psychology, 12*(1), 9–33.

Ballenger, C., and Rosebery, A. 2003. What counts as teacher research? Investigating the scientific and mathematical ideas of children from culturally diverse backgrounds. *Teachers College Record, 105*(2), 297–314.

Baltz, J. M., Katz, D. F., and Cone, R. A. 1988. Mechanics of sperm-egg interaction at the zona pellucida. *Biophysical Journal, 54*(4), 643–654.

Bang, M., and Medin, D. 2010. Cultural processes in science education: Supporting the navigation of multiple epistemologies. *Science Education, 94*(6), 1008–1026.

Bang, M., Medin, D. L., and Atran, S. 2007. Cultural mosaics and mental models of nature. *Proceedings of the National Academy of Sciences of the United States of America, 104*(35), 13868–13874.

Bang, M., Medin, D., Washinawatok, K., and Chapman, S. 2010. Innovations in culturally-based science education through partnerships and community. In M. S. Khine and I. M. Saleh (Eds.), *New science of learning: Cognition, computers and collaboration in education*, 569–592. New York: Springer.

Bang, M., B. Warren, A. S. Rosebery, and D. Medin. 2012. Desettling expectations in science education. *Human Development 55*, 302–318.

Barke, R. P., Jenkins-Smith, H., and Slovic, P. 1997. Risk perceptions of men and women scientists. *Social Science Quarterly, 78*(1), 167–176.

Barnhardt, R., and Kawagley, A. 2005. Indigenous knowledge systems and Alaska Native ways of knowing. *Anthropology and Education Quarterly, 36*(1), 8–23.

Basten, U., Biele, G., Heekeren, H. R., and Fiebach, C. J. 2010. How the brain integrates costs and benefits during decision making. *Proceedings of the National Academy of Sciences of the United States of America, 107*(50), 21767–21772.

Battiste, M. 2002. *Indigenous knowledge and pedagogy in First Nations education: A literature review with recommendations.* Ottowa: Indian and Northern Affairs Canada.

Battiste, M. 2008. The struggle and renaissance of Indigenous knowledge in Eurocentric education. In M. Villegas, S. R. Neugebauer, and K. R. Venegas (Eds.), *Indigenous knowledge and education: Sites of struggle, strength, and survivance,* 85–91. Cambridge, MA: Harvard Education Press.

Battiste, M., Bell, L., and Findlay, L. 2002. Decolonizing education in Canadian universities: An interdisciplinary, international, indigenous research project. *Canadian Journal of Native Education, 26*(2), 82–95.

Baxter, L. R., Jr., Schwartz, J. M., Bergman, K. S., Szuba, M. P., Guze, B. H., Mazziotta, J. C., et al. 1992. Caudate glucose metabolic rate changes with both drug and behavior therapy for obsessive-compulsive disorder. *Archives of General Psychiatry, 49*(9), 681–689.

Bazerman, M. H., and Neale, M. A. 1992. *Negotiating rationally.* New York: Free Press.

Beck, D. 2002. *Siege and survival: History of the Menominee Indians, 1634–1856.* Lincoln: University of Nebraska Press.

Beck, D. 2005. *The struggle for self-determination: History of the Menominee Indians since 1854.* Lincoln: University of Nebraska Press.

Bednar, J., and Page, S. 2007. Can game(s) theory explain culture? *Rationality and Society, 19*(1), 65–97.

Bell, P., Lewenstein, B., Shouse, A., and Feder, M. (Eds.) 2009. *Learning science in informal environments: People, places and pursuits.* Washington, DC: National Academies Press.

Bender, M. 2002. *Signs of Cherokee culture: Sequoyah's syllabary in Eastern Cherokee life.* Chapel Hill: University of North Carolina Press.

Bennis, W. M., Medin, D. L., and Bartels, D. M. 2010. The costs and benefits of calculation and moral rules. *Perspectives on Psychological Science, 5*(2), 187–202.

Berkes, F. 2008. *Sacred ecology.* New York: Routledge.

Berlin, B. 1992. *Ethnobiological classification: Principles of categorization of plants and animals in traditional societies*. Princeton: Princeton University Press.

Berlin, B., Berlin, E. A., Fernández Ugalde, J. C., Garcia Barrios, B. L., Puett, D., Nash, R., and González-Espinoza, M. 1999. The Maya ICBG: Drug discovery, medical ethnobiology, and alternative forms of economic development in the Highland Maya Region of Chiapas, Mexico. [Formerly *International Journal of Pharmacognosy*]. *Pharmaceutical Biology, 37*(Supplement 1), 127–144.

Berlin, E. A., Berlin, B., Lozoya, X., Meckes, M., Tortoriello, J., Villarreal, M., and Nader, L. 1996. The scientific basis of gastrointestinal herbal medicine among the highland Maya of Chiapas Mexico. In L. Nader (Ed.), *Naked science: Anthropological inquiry into boundaries, power, and knowledge*. London: Routledge.

Bleier, R. 1984. *Science and gender: A critique of biology and its theories on women*. Oxford: Pergamon Press.

Bloome, D. 2005. *Discourse analysis and the study of classroom language and literacy events: A microethnographic perspective*. Mahwah, NJ: Lawrence Erlbaum Associates.

Borkovec, T., Ray, W. J., and Stober, J. 1998. Worry: A cognitive phenomenon intimately linked to affective, physiological, and interpersonal behavioral processes. *Cognitive Therapy and Research, 22*(6), 561–576.

Borofsky, R., Barth, F., Shweder, R. A., Rodseth, L., and Stolzenberg, N. M. 2001. When: A conversation about culture. *American Anthropologist, 103*(2), 432–446.

Boster, J. S. 2005. Emotion categories across languages. In H. Cohen and C. Lefebvre (Eds.), *Handbook of categorization in cognitive science*, 187–222. Oxford: Elsevier.

Bowdle, B. F., and Gentner, D. 1997. Informativity and asymmetry in comparisons. *Cognitive Psychology, 34*(3), 244–286.

Bowdle, B. F., and Gentner, D. 2005. The career of metaphor. *Psychological Review, 112*(1), 193–216.

Bransford, J., and Brown, A. L. (Eds.). 2000. *How people learn: Brain, mind, experience, and school*. Expanded ed. Washington, DC: Commission on Behavioral and Social Sciences and Education of the National Research Council.

Brayboy, B. M. K. J. 2005. Toward a tribal critical race theory in education. *Urban Review, 37*(5), 425–446.

Bredo, E. 2009. Comments on Howe: Getting over the methodology wars. *Educational Researcher, 38*(6), 441–448. doi:10.3102/0013189x09343607.

Brooks, D. 2011. Social animal: How the new sciences of human nature can help make sense of a life. *New Yorker, 86*(44), 26–32.

Brumann, C., Abu-Lughod, L., Cerroni-Long, E., D'Andrade, R., Gingrich, A., Hannerz, U., et al. 1999. Writing for culture: Why a successful concept should not be discarded. *Current Anthropology, 40,* 1–27.

Buckwalter, W., and Stich, S. 2010. Gender and philosophical intuition. Social Science Research Network, Working Paper Series.

Burkhart, B. Y. 2004. What Coyote and Thales can teach us: An outline of American Indian epistemology. In A. Waters (Ed.), *American Indian thought: Philosophical essays,* 15–26. Malden, MA: Blackwell.

Burnett, R. C., Medin, D. L., Ross, N. O., and Blok, S. V. 2005. Ideal is typical. *Canadian Journal of Experimental Psychology/Revue canadienne de psychologie experimentale,* 59(1), 3–10.

Burton, L. 1995. Moving towards a feminist epistemology of mathematics. *Educational Studies in Mathematics, 28*(3), 275–291.

Cajete, G. 1994. *Look to the mountain: An ecology of indigenous education.* Durango, CO: Kivaki Press.

Cajete, G. 1999. *Igniting the sparkle: An indigenous science education model.* Skyland, NC: Kivaki Press.

Cajete, G. 2000. Indigenous knowledge: The Pueblo metaphor of Indigenous education. In M. Battiste (Ed.), *Reclaiming Indigenous voice and vision,* 181–191.Vancouver: University of British Columbia Press.

Cajete, G. 2004. Philosophy of native science. In A. Waters (Ed.), *American Indian thought: Philosophical essays,* 45–57. Malden, MA: Blackwell.

Cameron, J. 2006. *Rewards and intrinsic motivation: Resolving the controversy.* Charlotte, NC: Information Age Publishing.

Carey, S. 1985. *Conceptual change in childhood.* Cambridge, MA: MIT Press.

Carey, S. 2009. *The origin of concepts.* New York: Oxford University Press.

Carey, S., and Spelke, E. 1994. Domain-specific knowledge and conceptual change. In L. A. Hirschfeld and S. A. Gelman (Eds.), *Mapping the mind: Domain-specificity in culture and cognition,* 169–200. New York: Cambridge University Press.

Carroll, K. M., and Onken, L. S. 2005. Behavioral therapies for drug abuse. *American Journal of Psychiatry, 162*(8), 1452–1460.

Chandler, M. J., and Lalonde, C. 1998. Cultural continuity as a hedge against suicide in Canada's First Nations. *Transcultural Psychiatry, 35*(2), 191–219.

Chang, H. 2008. The myth of the boiling point. *Science Progress, 91*(3), 219–240.

Chavajay, P., and Rogoff, B. 1999. Cultural variation in management of attention by children and their caregivers. *Developmental Psychology*, *35*(4), 1079–1090.

Cole, M. 1996. *Culture in mind*. Cambridge, MA: Harvard University Press.

Cole, M., and Scribner, S. 1974. *Culture and thought: A psychological introduction*. New York: Wiley.

Coley, J. D., Medin, D. L., and Atran, S. 1997. Does rank have its privilege? Inductive inferences within folkbiological taxonomies. *Cognition*, *64*(1), 73–112.

Cook, H. M. 1999. Language socialization in Japanese elementary schools: Attentive listening and reaction turns. *Journal of Pragmatics*, *31*(11), 1443–1465.

Correa-Chávez, M., Rogoff, B., and Mejia Arauz, R. 2005. Cultural patterns in attending to two events at once. *Child Development*, *76*(3), 664–678.

Cosmides, L., Tooby, J., and Barkow, J. H. 1992. *The adapted mind*. New York: Oxford University. Press.

Dake, K. 1992. Myths of nature: Culture and the social construction of risk. *Journal of Social Issues*, *48*(4), 21–37.

Darwin, C. R. [1872] 1998. *The origin of species and the descent of man*. New York: Modern Library.

Daston, L., and Lunbeck, E. 2011. *Histories of scientific observation*. Chicago: University of Chicago Press.

Davis, T. 2000. *Sustaining the forest, the people, and the spirit*. Albany: State University of New York Press.

Dehghani, M., Bang, M., Medin, D. L., Marin, A., Leddon, E., and Waxman, S. R. 2013. Epistemologies in text in children's books: Native and non-native authored books. *International Journal of Science Education*.

Deloria, P. J. 1998. *Playing Indian*. New Haven: Yale University Press.

Deloria, V., Jr. 1998. *For this land: Writings on religion in America*. New York: Routledge.

Deloria, V., and Wildcat, D. 2001. *Power and place*. Golden, CO: Fulcrum Resources.

Demmert, W. G., Jr., and Towner. J. C. 2003. A review of the research literature on the influences of culturally based education on the academic performance of Native American students. Portland, OR, Northwest Regional Educational Laboratory. Retrieved from http://educationnorthwest.org/resource/561 on May 30, 2013.

Diamond, J. 1997. *Guns, germs, and steel: The fates of human societies*. New York: W. W. Norton.

Di Chiro, G. 2004. Local actions, global visions. In R. Eglash, J. L. Croissant, G. Di Chiro, and R. Fouché (Eds.), *Appropriating technology: Vernacular science and social power*, 225–252. Minneapolis: University of Minnesota Press.

Douglas, M., and Wildavsky, A. 1983. *Risk and culture: An essay on the selection of technical and environmental dangers*. Berkeley: University of California Press.

Driver, R., Newton, P., and Osborne, J. 2000. Establishing the norms of scientific argumentation in classrooms. *Science Education, 84*(3), 287–312.

Duster, T. 1996. The prism of heritability and the sociology of knowledge. In L. Nader (Ed.), *Naked science: Anthropological inquiry into boundaries, power, and knowledge*, 119–130. London: Routledge.

Eagly, A. H. 2005. Achieving relational authenticity in leadership: Does gender matter? *Leadership Quarterly, 16*(3), 459–474.

Eberbach, C., and Crowley, K. 2009. From everyday to scientific observation: How children learn to observe the biologist's world. *Review of Educational Research, 79*(1), 39–68.

Edelson, D. C. 2002. Design research: What we learn when we engage in design. *Journal of the Learning Sciences, 11*(1), 105–121.

Eisenberger, R., and Cameron, J. 1996. Detrimental effects of reward: Reality or myth? *American Psychologist, 51*(11), 1153–1166.

Epley, N., Waytz, A., and Cacioppo, J. T. 2007. On seeing human: A three-factor theory of anthropomorphism. *Psychological Review, 114*(4), 864–886.

Erduran, S., Simon, S., and Osborne, J. 2004. TAPping into argumentation: Developments in the application of Toulmin's argument pattern for studying science discourse. *Science Education, 88*(6), 915–933.

Evans-Pritchard, E. E. 1937. *Witchcraft, magic and oracles among the Azande*. Oxford: Clarendon Press.

Finucane, M. L., Slovic, P., Mertz, C. K., Flynn, J., and Satterfield, T. A. 2000. Gender, race, and perceived risk: the "white male" effect. *Health Risk and Society, 2*(2), 159–172.

Fischhoff, B., Gonzalez, R. M., Small, D. A., and Lerner, J. S. 2003. Judged terror risk and proximity to the World Trade Center. *Journal of Risk and Uncertainty, 26*(2), 137–151.

Flood, R. S. 1995. *Lost bird of Wounded Knee: Spirit of the Lakota*. New York: Perseus Books.

Foreman, G. 1938. *Sequoyah*. Norman: University of Oklahoma Press.

Frazer, J. G. 1890, 1912. *The golden bough: A study in comparative religion*. 2 vols. London: Macmillan.

Frisch, J. E. 1963. Sex-differences in the canines of the gibbon (*Hylobates lar*). *Primates*, 4(2), 1–10.

Frodeman, R., and Holbrook, J. B. 2011. NSF's struggle to articulate relevance. *Science*, 333(6039), 157–158.

Galinsky, A. D., Magee, J. C., Inesi, M. E., and Gruenfeld, D. H. 2006. Power and perspectives not taken. *Psychological Science*, 17(12), 1068–1074.

Gardipee, F. M., Strobel, D. A., Allendorf, F. W., Luikart, G., Hebblewhite, M., and Clow, R. 2007. Development of fecal DNA sampling methods to assess genetic population structure of greater Yellowstone bison. Master's thesis, University of Montana, Missoula.

Garrett, P. B., and Baquedano-López, P. 2002. Language socialization: Reproduction and continuity, transformation and change. *Annual Review of Anthropology*, 31, 339–361.

Gastil, J., Braman, D., Kahan, D., and Slovic, P. 2011. The cultural orientation of mass political opinion. *PS: Political Science and Politics*, 44(04), 711–714.

Gelman, R. 1990. First principles organize attention to and learning about relevant data: Number and the animate-inanimate distinction as examples. *Cognitive Science*, 14(1), 79–106.

Gelman, R., and Brenneman, K. 2004. Science learning pathways for young children. *Early Childhood Research Quarterly*, 19(1), 150–158.

Gelman, R., Brenneman, K., Macdonald, G., and Roman, M. 2010. *Preschool pathways to science (PrePS): Facilitating scientific ways of thinking, talking, doing, and understanding*. Baltimore: Brookes Publishing.

Gelman, S. A. 2003. *The essential child: Origins of essentialism in everyday thought*. New York: Oxford University Press.

Gentner, D., and Goldin-Meadow, S. 2003. *Language in Mind*. Cambridge, MA: MIT Press.

Gentner, D., and Markman, A. B. 1997. Structure mapping in analogy and similarity. *American Psychologist*, 52(1), 45–56.

Ghiselin, M. T. 1981. Categories, life, and thinking. *Behavioral and Brain Sciences*, 4(02), 269–283.

Gielen, U., and Kelly, D. 1983. *Buddhist Ladakh: Psychological portrait of a non-violent culture*. Paper presented at the Annual Conference of the Society for Cross-Cultural Research, Washington, DC.

Giere, R. N. 2006. Perspectival pluralism. In S. H. Kellert, H. E. Longino, and C. K. Waters (Eds.), *Scientific pluralism*, 26–41. Minneapolis: University of Minnesota Press.

Gilligan, C. 1977. In a different voice: Women's conceptions of self and of morality. *Harvard Educational Review*, *47*(4), 481–517.

Gilligan, C. 1982. *In a different voice: Psychological theory and women's development*. Cambridge, MA: Harvard University Press.

Gleitman, L. R., Gleitman, H., Miller, C., and Ostrin, R. 1996. Similar, and similar concepts. *Cognition*, *58*(3), 321–376.

Godfrey-Smith, P. 2006. The strategy of model-based science. *Biology and Philosophy*, *21*(5), 725–740.

Goldapple, K., Segal, Z., Garson, C., Lau, M., Bieling, P., Kennedy, S., et al. 2004. Modulation of cortical-limbic pathways in major depression: Treatment-specific effects of cognitive behavior therapy. *Archives of General Psychiatry*, *61*(1), 34–41.

Goodwin, C. 1984. Notes on story structure and the organization of participation. In J. M. Atkinson and J. Heritage (Eds.), *Structures of social action*, 225–246. New York: Cambridge University Press.

Gopnik, A., Glymour, C., Sobel, D. M., Schulz, L. E., Kushnir, T., and Danks, D. 2004. A theory of causal learning in children: Causal maps and Bayes nets. *Psychological Review*, *111*(1), 3–32.

Gopnik, A., and Schulz, L. 2004. Mechanisms of theory formation in young children. *Trends in Cognitive Sciences*, *8*(8), 371–377.

Gopnik, A., and Schulz, L. 2007. *Causal learning: Psychology, philosophy, and computation*. New York: Oxford University Press.

Gowaty, P. A. (Ed.) 1997. *Feminism and evolutionary biology*. New York: International Thompson Publishing.

Grande, S. M. A. 2000. American Indian geographies of identity and power: At the crossroads of indigena and mestizaje. *Harvard Educational Review*, *70*(4), 467–499.

Grandin, T., and Johnson, C. 2005. *Animals in translation: Using the mysteries of autism to decode animal behavior*. New York: Scribner.

Griffiths, T. L., and Tenenbaum, J. B. 2005. Structure and strength in causal induction. *Cognitive Psychology*, *51*(4), 334–384.

Grigg, W. S., Lauko, M. A., and Brockway, D. M. 2006. *The nation's report card: Science 2005*. Washington, DC: U.S. Department of Education.

Grignon, D., Alegria, R., Dodge, C., Lyons, G., Waukechon, C., Warrington, C., et al. 1998. *Menominee tribal history guide: Commemorating Wisconsin sesquicentennial 1848–1998*. Keshena: Menominee Indian Tribe of Wisconsin.

Gross, M., and Averill, M. 2003. Evolution and patriarchal myths of scarcity and competition. In S. Harding and M. Hintikka (Eds.), *Discovering reality: Feminist perspectives on epistemology, metaphysics, methodology and philosophy of science*, 2nd ed., 71–95. Dordrecht: Kluwer Academic Publishers.

Gutheil, G., Vera, A., and Keil, F. C. 1998. Do houseflies think? Patterns of induction and biological beliefs in development. *Cognition, 66*(1), 33–49.

Gutiérrez, K. D. 2006. *Culture matters: Rethinking educational equity.* New York: Carnegie Foundation.

Gutiérrez, K. D., Baquedano-López, P., and Asato, J. 2000. "English for the children": The new literacy of the old world order, language policy and educational reform. *Bilingual Research Journal, 24*(1–2), 87–112.

Gutiérrez, K. D., and Rogoff, B. 2003. Cultural ways of learning: Individual traits or repertoires of practice. *Educational Researcher, 32*(5), 19–25.

Guyette, S. 1983. *Community-based research: A handbook for Native Americans.* Los Angeles: American Indian Studies Center, University of California.

Hacking, I. 1992. Statistical language, statistical truth and statistical reason: The self-authentification of a style of scientific reasoning. In E. McMullin (Ed.), *The social dimensions of science,*3: 130–157. Notre Dame, IN: University of Notre Dame Press.

Hall, E. T. 1976. *Beyond culture.* Garden City, NY: Anchor.

Hall, P., and Pecore, M. 1995. *Case study: Menominee tribal enterprises.* Madison: Institute for Environmental Studies and the Land Tenure Center, University of Wisconsin-Madison.

Hammer, D., and Elby, A. 2003. Tapping epistemological resources for learning physics. *Journal of the Learning Sciences, 12*(1), 53–90.

Haraway, D. J. 1988. Situated knowledges: The science question in feminism and the privilege of partial perspective. *Feminist Studies, 14*(3), 575–599.

Haraway, D. J. 1989. *Primate visions.* New York: Routledge.

Harding, S. 1994. Is science multicultural? Challenges, resources, opportunities, uncertainties. *Configurations, 2*(2), 301–330.

Harding, S. 2006. *Science and social inequality: Feminist and postcolonial issues.* Champaign: University of Illinois Press.

Harding, S. 2008. *Sciences from below: Feminisms, postcolonialities, and modernities.* Durham: Duke University Press Books.

Hatano, G., and Inagaki, K. 2000. Domain-specific constraints of conceptual development. *International Journal of Behavioral Development, 24*(3), 267–275.

Haury, D. L. 2002. *Fundamental skills in science: Observation*. Columbus, OH: ERIC Clearinghouse for Science Mathematics and Environmental Education.

Heart, M. Y. H. B., and DeBruyn, L. M. 1998. The American Indian holocaust: Healing historical unresolved grief. *American Indian and Alaska Native Mental Health Research, 8*(2), 60–82.

Heath, C. 1995. Escalation and de-escalation of commitment in response to sunk costs: The role of budgeting in mental accounting. *Organizational Behavior and Human Decision Processes, 62*(1), 38–54.

Heiman, M. K. 1997. Science by the people: Grassroots environmental monitoring and the debate over scientific expertise. *Journal of Planning Education and Research, 16*(4), 291–299.

Henderson, M. D., Fujita, K., Trope, Y., and Liberman, N. 2006. Transcending the "here": The effect of spatial distance on social judgment. *Journal of Personality and Social Psychology, 91*(5), 845–856.

Henrich, J., S. J. Heine, and A. Norenzayan. 2010. Most people are not WEIRD. *Nature* 466 (7302): 29.

Henriksen, G. 2009. *I dreamed the animals: Kaniuekutat: The life of an Innu hunter*. Oxford, NY: Berghahn Books.

Hermes, M. 1999. Research methods as a situated response: Toward a First Nations' methodology. In L. Parker, D. Deyhle, and S. Villenas (Eds.), *Race is . . . race isn't: Critical race theory and qualitative studies in education*, 83–100. Boulder, CO: Westview Press.

Hermes, M. 2000. The scientific method, Nintendo, and Eagle feathers: Rethinking the meaning of "culture-based" curriculum at an Ojibwe tribal school. *International Journal of Qualitative Studies in Education, 13*(4), 387–400.

Herrmann, P., Waxman, S. R., and Medin, D. L. 2010. Anthropocentrism is not the first step in children's reasoning about the natural world. *Proceedings of the National Academy of Sciences of the United States of America, 107*(22), 9979–9984.

Hess, D. J. 1995. *Science and technology in a multicultural world: The cultural politics of facts and artifacts*. New York: Columbia University Press.

Hickling, A. K., and Gelman, S. A. 1995. How does your garden grow? Early conceptualization of seeds and their place in the plant growth cycle. *Child Development, 66*(3), 856–876.

Hirschfeld, L. A. 2002. Why don't anthropologists like children? *American Anthropologist, 104*(2), 611–627.

Hirschfeld, L. A., and Gelman, S. A. 1994. *Mapping the mind: Domain-specificity in culture and cognition*. New York: Cambridge University Press.

Hobson, J. M. 2004. *The Eastern origins of Western civilisation*. Cambridge: Cambridge University Press.

Hosmer, B. C. 1999. *American Indians in the marketplace: Persistence and innovation among the Menominees and Metlakatlans, 1870–1920*. Lawrence: University Press of Kansas.

Howe, K. R. 2009. Positivist dogmas, rhetoric, and the education science question. *Educational Researcher, 38*(6), 428–440.

Hubbard, B. M. 2003. Conscious evolution: The next stage of human development. *Systems Research and Behavioral Science, 20*(4), 359–370.

Hubbard, R. 1979. Have only men evolved? In S. Harding and M. Hintikka (Eds.), *Discovering reality*, 45–70. Cambridge, MA: Schenkman Publishing.

Hull, D. L. 1988. *Science as a process: An evolutionary account of the social and conceptual development of science*. Chicago: University of Chicago Press.

Iliev, R. 2012. Cultural ties. Association for Psychological Science video, 30 March. http://www.psychologicalscience.org/index.php/video/cultural-ties.html.

Inagaki, K. 1990. The effects of raising animals on children's biological knowledge. *British Journal of Developmental Psychology, 8*(2), 119–129.

Inagaki, K., and Hatano, G. 1993. Young children's understanding of the mind-body distinction. *Child Development, 64*(5), 1534–1549.

Inagaki, K., and Hatano, G. 2002. *Young children's naive thinking about the biological world*. London: Psychology Press.

Ingold, T. 2000. *The perception of the environment: Essays on livelihood, dwelling and skill*. London: Psychology Press.

Insel, T. 2010. Faulty circuits. *Scientific American Magazine, 302*(4), 44–51.

Itani, J. 1958. On the acquisition and propagation of a new food habit in the natural group of the Japanese monkey at Takasaki-Yama. *Primates, 1*(2), 84–98.

Itani, J. 1961. The society of Japanese Monkeys. *Japan Quarterly (Asahi Shinbunsha), 8*(4), 421–430.

Johnson, R. B. 2009. Comments on Howe: Toward a more inclusive "scientific research in education." *Educational Researcher, 38*(6), 449–457. doi:10.3102/0013189x09344429.

Kahan, D. M., Braman, D., Gastil, J., Slovic, P., and Mertz, C. 2007. Culture and identity-protective cognition: Explaining the white-male effect in risk perception. *Journal of Empirical Legal Studies, 4*(3), 465–505.

Kahan, D. M., Jenkins-Smith, H., and Braman, D. 2011. Cultural cognition of scientific consensus. *Journal of Risk Research, 14*(2), 147–174.

Kahan, D. M., Slovic, P., Braman, D., and Gastil, J. 2006. Fear of democracy: A cultural evaluation of Sunstein on risk. *Harvard Law Review, 119*(4), 1071–1109.

Kahn, P. H., and Kellert, S. R. 2002. *Children and nature: Psychological, sociocultural, and evolutionary investigations.* Cambridge, MA: MIT Press.

Kahneman, D., and Tversky, A. 1979. Prospect theory: An analysis of decision under risk. *Econometrica, 47*(2), 263–291.

Kawagley, A. O. 1995. *A Yupiaq worldview: A pathway to ecology and spirit.* Prospect Heights, IL: Waveland Press.

Kawai, M. 1965. Newly-acquired pre-cultural behavior of the natural troop of Japanese monkeys on Koshima Islet. *Primates, 6*(1), 1–30.

Keil, F. C. 1981. Constraints on knowledge and cognitive development. *Psychological Review, 88*(3), 197–227.

Keil, F. C. 1992. The origins of an autonomous biology. In M. R. Gunnar and M. Maratsos (Eds.), *Modularity and constraints in language and cognition,* 103–137. Hillsdale, NJ: Lawrence Erlbaum Associates.

Keller, E. F. 1984. *A feeling for the organism: The life and work of Barbara McClintock.* New York: Henry Holt.

Keller, E. F. 1986. How gender matters, or, why it's so hard for us to count past two. In J. Harding (Ed.), *Perspectives on gender and science,* 168–183. London: Palmer Press.

Keller, E. F. 1992. *Secrets of life, secrets of death: Essays on language, gender, and science.* London: Psychology Press.

Keller, E. F. 2010. *The mirage of a space between nature and nurture.* Durham: Duke University Press.

Keller, E. F., and Longino, H. E. 1996. *Feminism and science.* New York: Oxford University Press.

Keller, E. F. 1984. *A feeling for the organism: The life and work of Barbara McClintok.* New York: Henry Holt.

Kellert, S. R. 1993. Values and perceptions of invertebrates. *Conservation Biology, 7*(4), 845–855.

Kellert, S. R. 1996. *The value of life: Biological diversity and human society.* Washington, DC: Island Press.

Kellert, S. R. 1997. Environmental values, the coastal context, and a sense of place. In L. A. Brooks and S. D. VanDeVeer (Eds.), *Saving the seas: Values, scientists, and international governance*, 47–66. College Park, MD: Maryland Sea Grant.

Kellman, P. J., and Spelke, E. S. 1983. Perception of partly occluded objects in infancy. *Cognitive Psychology, 15*(4), 483–524.

King, P. M., and Kitchener, K. S. 1994. *Developing reflective judgment: Understanding and promoting intellectual growth and critical thinking in adolescents and adults*. San Francisco: Jossey-Bass.

Knorr-Cetina, K. 1981. *The manufacture of knowledge*. Oxford: Pergamon Press.

Kohlberg, L. 1973. The claim to moral adequacy of a highest stage of moral judgment. *Journal of Philosophy, 70*(18), 630–646.

Kohlberg, L. 1984. *The psychology of moral development: The nature and validity of moral stages*. San Francisco: Harper and Row.

Korzybski, A. 1958. *Science and sanity: An introduction to non-Aristotelian systems and general semantics*. New York: Institute of General Semantics.

Kovach, M. 2010. Conversational method in indigenous research. *First Peoples Child and Family Review, 5*(1), 40–48.

Kraus, N., Malmfors, T., and Slovic, P. 1992. Intuitive toxicology: Expert and lay judgments of chemical risks. *Risk Analysis, 12*(2), 215–232.

Kühberger, A. 1998. The influence of framing on risky decisions: A meta-analysis. *Organizational Behavior and Human Decision Processes, 75*(1), 23–55.

Kuhn, T. S. 1962. *The structure of scientific revolutions*. Chicago: University of Chicago Press.

Kuhn Berland, L. 2008. *Understanding the composite practice that forms when classrooms take up the practice of scientific argumentation*. Ann Arbor: Proquest, LLC.

Lakoff, G., and Johnson, M. 1980. *Metaphors we live by*. Chicago: University of Chicago Press.

Larrick, R. P., Morgan, J. N., and Nisbett, R. E. 1990. Teaching the use of cost-benefit reasoning in everyday life. *Psychological Science, 1*(6), 362–370.

Leach, E. R. 1957. The epistemological background to Malinowski's empiricism. In R. Firth (Ed.), *Man and culture: An evaluation of the work of Bronislaw Malinowski*, 119–137. London: Routledge and Kegan Paul.

Lederman, N. G., Abd-El-Khalick, F., Bell, R. L., and Schwartz, R. S. 2002. Views of nature of science questionnaire: Toward valid and meaningful assessment of learners' conceptions of nature of science. *Journal of Research in Science Teaching, 39*(6), 497–521.

Ledward, B., Takayama, B., and Kahumoku, W. III. 2008. *Kiki Na Wai: Swiftly flowing streams. Examples of 'Ohana and community integration in culture-based education.* Honolulu: Kamehameha Schools, Research and Evaluation Division.

Lee, C. D. 1995. A culturally based cognitive apprenticeship: Teaching African American high school students skills in literary interpretation. *Reading Research Quarterly, 30*(4), 608–630.

Lee, C. D. 2001. Is October Brown Chinese? A cultural modeling activity system for underachieving students. *American Educational Research Journal, 38*(1), 97–141.

Lee, C. D., Spencer, M. B., and Harpalani, V. 2003. "Every shut eye ain't sleep": Studying how people live culturally. *Educational Researcher, 32*(5), 6–13.

Lee, O., and Fradd. S. H. 1998. Science for all, including students from non-English-language backgrounds. *Educational Researcher, 27*(4), 12–21.

Lee, S. T., and Lin, H. S. 2005. Using argumentation to investigate science teachers' teaching practices: The perspective of instructional decisions and justifications. *International Journal of Science and Mathematics Education, 3*(3), 429–461.

Leggett, A. J. 1966. Notes on the writing of scientific English for Japanese physicists. *Journal of the Physical Society of Japan, 21*, 790–805.

Le Guen, O. 2011. Speech and gesture in spatial language and cognition among the Yucatec Mayas. *Cognitive Science, 35*(5), 905–938.

Lehman, D. R., and Nisbett, R. E. 1990. A longitudinal study of the effects of undergraduate training on reasoning. *Developmental Psychology, 26*(6), 952–960.

Lei, T., and Cheng, S. W. 1984. An empirical study of Kohlberg's theory and scoring system of moral judgment in Chinese society. Unpublished manuscript, Harvard University, Center for Moral Education, Cambridge, MA.

Lemke, J. L. 1990. *Talking science: Language, learning, and values.* New York: Ablex Publishing.

Lepper, M. R., Greene, A., and Nisbett, R. E. 1973. Undermining children's intrinsic interest with extrinsic reward: A test of the "overjustification" hypothesis. *Journal of Personality and Social Psychology, 28*(1), 129–137.

Leshner, A. 2011. We need to reward those who nurture a diversity of ideas in science. *Chronicle of Higher Education,* March 6. http://chronicle.com/article/We-Need-to-Reward-Those-Who/126591/

Levin, I. P., and Chapman, D. P. 1993. Risky decision making and allocation of resources for leukemia and AIDS programs. *Health Psychology, 12*(2), 110.

Levine, G. 1996. What is science studies for and who cares? In A. Ross (Ed.), *Science wars,* 123–138. Durham: Duke Univeristy Press.

Levins, R., and Lewontin, R. C. 1985. *The dialectical biologist*. Cambridge, MA: Harvard University Press.

Lewontin, R. C. 1996. A la recherche du temps perdu: A review essay. In A. Ross (Ed.), *Science wars*, 293–301. Durham: Duke University Press.

Liberman, N., and Trope, Y. 2008. The psychology of transcending the here and now. *Science*, *322*(5905), 1201–1205.

Liberman, N., Trope, Y., McCrea, S. M., and Sherman, S. J. 2007. The effect of level of construal on the temporal distance of activity enactment. *Journal of Experimental Social Psychology*, *43*(1), 143–149.

Lima, M. L., and Castro, P. 2005. Cultural theory meets the community: Worldviews and local issues. *Journal of Environmental Psychology*, *25*(1), 23–35.

Lindenberg, S. 2001. Intrinsic motivation in a new light. *Kyklos*, *54*(2–3), 317–342.

Lippman, A. 1991. Prenatal genetic testing and screening: Constructing needs and reinforcing inequities. *American Journal of Law and Medicine*, *17*, 15–50.

Lloyd, E. A. 1993. Pre-theoretical assumptions in evolutionary explanations of female sexuality. *Philosophical Studies*, *69*(2), 139–153.

Lloyd, G. 2010. History and human nature: Cross-cultural universals and cultural relativities. *Interdisciplinary Science Reviews*, *35*(3–4), 201–214.

Lomawaima, K. T. 1995. *They called it prairie light: The story of Chilocco Indian school*. Lincoln: University of Nebraska Press.

Lomawaima, K. T. 2000. Tribal sovereigns: Reframing research in American Indian education. *Harvard Educational Review*, *70*(1), 1–23.

Lombrozo, T. 2010. Causal-explanatory pluralism: How intentions, functions, and mechanisms influence causal ascriptions. *Cognitive Psychology*, *61*(4), 303–332.

Longino, H. E. 1987. Can there be a feminist science? *Hypatia*, *2*(3), 51–64.

Longino, H. E. 1990. *Science as social knowledge: Values and objectivity in scientific inquiry*. Princeton: Princeton University Press.

Longino, H. E. 2002. *The fate of knowledge*. Princeton: Princeton University Press.

Longino, H. E. 2006. Theoretical pluralism and the scientific study of behavior. In *Scientific pluralism*. Minneapolis: University of Minnesota Press.

López, A., Atran, S., Coley, J. D., Medin, D. L., and Smith, E. E. 1997. The tree of life: Universal and cultural features of folkbiological taxonomies and inductions. *Cognitive Psychology*, *32*, 251–295.

Louv, R. 2008. *Last child in the woods: Saving our children from nature-deficit disorder*. Chapel Hill, NC: Algonquin Books.

Luhrmann, T. M. 2011. Toward an anthropological theory of mind. *Suomen Antropologi: Journal of the Finnish Anthropological Society*, *36*(3), 5–69.

Luo, Y., Kaufman, L., and Baillargeon, R. 2009. Young infants' reasoning about physical events involving inert and self-propelled objects. *Cognitive Psychology*, *58*(4), 441–486.

Malinowski, B. 1925. *Magic, science and religion; and other essays*. Garden City, NY: Doubleday Anchor Books.

Malt, B. C. 1995. Category coherence in cross-cultural perspective. *Cognitive Psychology*, *29*(2), 85–148.

Marin, A., and Bang, M. In revision. Repatriating science teaching and learning: Finding our way to storywork. *Journal of American Indian Education*.

Marr, D. 1982. *Vision: A computational investigation into the human representation and processing of visual information*. New York: Henry Holt.

Martin, E. 1996. The egg and the sperm: How science has constructed a romance based on stereotypical male-female roles. In E. F. Keller and H. E. Longino (Eds.), *Feminism and science*. New York: Oxford University Press.

Massey, C. M., and Gelman, R. 1988. Preschooler's ability to decide whether a photographed unfamiliar object can move itself. *Developmental Psychology*, *24*(3), 307–317.

Masuda, T., and Nisbett, R. E. 2006. Culture and change blindness. *Cognitive Science*, *30*(2), 381–399.

Mayr, E. 1957. *Species concepts and definitions*. Washington, DC: AAAS.

McCabe, D. P., and Castel, A. D. 2008. Seeing is believing: The effect of brain images on judgments of scientific reasoning. *Cognition*, *107*(1), 343–352.

Medin, D. L., and Bang, M. 2008. Perspective taking, diversity and partnerships. *Psychological Science Agenda*, *22*(2). Retrieved from http://www.apa.org/science/about/psa/2008/02/index.aspx on May 30, 2013.

Medin, D. L., Goldstone, R. L., and Gentner, D. 1993. Respects for similarity. *Psychological Review*, *100*, 254–278.

Medin, D. L., Lynch, E. B., Coley, J. D., and Atran, S. 1997. Categorization and reasoning among tree experts: Do all roads lead to Rome? *Cognitive Psychology*, *32*, 49–96.

Medin, D. L., Lynch, E. B., and Solomon, K. O. 2000. Are there kinds of concepts? *Annual Review of Psychology*, *51*(1), 121–147.

Medin, D. L., Ross, N., Atran, S., Burnett, R. C., and Blok, S. V. 2002. Categorization and reasoning in relation to culture and expertise. *Psychology of Learning and Motivation*, *41*, 1–41.

Medin, D. L., Ross, N. O., Atran, S., Cox, D., Coley, J., Proffitt, J. B., et al. 2006. Folkbiology of freshwater fish. *Cognition*, *99*(3), 237–273.

Medin, D. L., and Waxman, S. R. 2007. Interpreting asymmetries of projection in children's inductive reasoning. In A. Feeney and E. Heit (Eds.), *Inductive reasoning*, 55–80. New York: Cambridge University Press.

Medin, D. L., Waxman, S. R., Woodring, J., and Washinawatok, K. 2010. Human-centeredness is not a universal feature of young children's reasoning: Culture and experience matter when reasoning about biological entities. *Cognitive Development*, *25*(3), 197–207.

Meriam, L. 1928. The problem of Indian administration. Report of a survey made at the request of Honorable Hubert Work, Secretary of the Interior, and submitted to him, February 21, 1928, Washington, DC: Brookings Institution.

Mervis, J. 2011. Beyond the data. *Science*, *334*(6053), 169–171.

Meyer, M. A. 1998. Native Hawaiian epistemology: Sites of empowerment and resistance. *Equity and Excellence*, *31*(1), 22–28.

Meyer, M. A. 2001. Our own liberation: Reflections on Hawaiian epistemology. *Contemporary Pacific*, *13*(1), 124–148.

Mihesuah, D. A. 1998. *Natives and academics: Researching and writing about American Indians*. Lincoln: University of Nebraska Press.

Miller, G. A. 2010. Mistreating psychology in the decades of the brain. *Perspectives on Psychological Science*, *5*(6), 716–743. doi:10.1177/1745691610388774.

Moje, E. B., Collazo, T., Carrillo, R., and Marx. R. W. 2001. "Maestro, what is 'quality'?": Language, literacy, and discourse in project based science. *Journal of Research in Science Teaching*, *38*(4), 469–498.

Morling, B., and Lamoreaux, M. 2008. Measuring culture outside the head: A meta-analysis of individualism—collectivism in cultural products. *Personality and Social Psychology Review*, *12*(3), 199–221.

Motokawa, T. 1989. Sushi science and hamburger science. *Perspectives in Biology and Medicine*, *32*(4), 489–504.

Murayama, K., Matsumoto, M., Izuma, K., and Matsumoto, K. 2010. Neural basis of the undermining effect of monetary reward on intrinsic motivation. *Proceedings of the National Academy of Sciences of the United States of America*, *107*(49), 20911–20916.

Nadasdy, P. 2005. *Hunters and bureaucrats: Power, knowledge, and aboriginal-state relations in the southwest Yukon.* Vancouver: University of British Columbia Press.

Nader, L. (Ed.). 1996. *Naked science: Anthropological inquiry into boundaries, power, and knowledge.* London: Routledge.

Nader, L. 2010. The three-cornered constellation: Magic, science, and religion revisited. In L. Nader (Ed.), *The energy reader,* 205–218. Chichester, UK: Blackwell.

Nasir, N. S., and Hand, V. M. 2006. Exploring sociocultural perspectives on race, culture, and learning. *Review of Educational Research, 76*(4), 449–475.

Nasir, N. S., and Saxe, G. B. 2003. Ethnic and academic identities: A cultural practice perspective on emerging tensions and their management in the lives of minority students. *Educational Researcher, 32*(5), 14–18.

National Academy of Sciences, National Academy of Engineering, and Institute of Medicine of the National Academies. 2007. *Rising above the gathering storm: Energizing and employing American for a brighter economic future.* Washington, DC: National Academies Press.

National Academy of Sciences, National Academy of Engineering, and Institute of Medicine of the National Academies. 2010. *Expanding underrepresented minority participation: America's science and technology talent at the crossroads.* Washington, DC: National Academies Press.

National Research Council. 2007. *Taking science to school: Learning and teaching science in grades K-8.* Washington, DC: National Academies Press.

National Research Council. 2009. *Learning science in informal environments: People, places, and pursuits.* Washington, DC: National Academies Press.

National Science Foundation (NSF). 2007. *Women, minorities, and persons with disabilities in science and engineering.* Arlington, VA: National Science Foundation.

Needham, J. 1959. *Science and civilisation in China: Mathematics and the sciences of the heavens and the earth.* Cambridge: Cambridge University Press.

Nemeroff, C., and Rozin, P. 1994. The contagion concept in adult thinking in the United States: Transmission of germs and of interpersonal influence. *Ethos* (Berkeley, Calif.), *22*(2), 158–186.

Newman, J. G. 1967. *The Menominee forest of Wisconsin: A case history in American forest management.* East Lansing: Michigan State University Department of Forestry.

Nisbett, R. E. 2003. *The geography of thought: How Asians and Westerners think differently—and why.* New York: Free Press.

Nisbett, R. E. 2009. *Intelligence and how to get it: Why schools and cultures count.* New York: W. W. Norton.

Nisbett, R. E., and Cohen, D. 1996. *Culture of honor: The psychology of violence in the South*. Boulder, CO: Westview Press.

Nisbett, R. E., and Masuda, T. 2003. Culture and point of view. *Proceedings of the National Academy of Sciences, 100*(19), 11163–11170.

Nisbett, R. E., Peng, K., Choi, I., and Norenzayan, A. 2001. Culture and systems of thought: Holistic versus analytic cognition. *Psychological Review, 108*(2), 291–310.

Noel, J. R. 2002. Education toward cultural shame: A century of Native American education. *Educational Foundations, 16*(1), 19–32.

Ochs, E., Gonzales, P., and Jacoby, S. 1996. "When I come down I'm in the domain state": Grammar and graphic representation in the interpretive activity of physicists. In E. Ochs, E. A. Schegloff, and S. A. Thompson (Eds.), *Interaction and grammar*, 328–369. New York: Cambridge University Press.

Ochs, E., Taylor, C., Rudolph, D., and Smith, R. 1992. Storytelling as a theory-building activity. *Discourse Processes, 15*(1), 37–72.

Olsen, S. 2006. *Yetsa's sweater*. J. Larson, illustrator. Winlaw, BC: Sono Nis Press.

Orellana, M. F., and D'warte, J. 2010. Recognizing different kinds of "head starts." *Educational Researcher, 39*(4), 295–300.

Oreskes, N., and Conway, E. M. 2010. *Merchants of doubt*. London: Bloomsbury.

Page, S. E. 2007. *The difference: How the power of diversity creates better groups, firms, schools, and societies*. Princeton: Princeton University Press.

Palmer, S. E., and Kimchi, R. 1986. The information processing approach to cognition. In T. J. Knapp and L. Robertson (Eds.), *Approaches to cognition: Contrasts and controversies*, 37–77. Hillsdale, NJ: Lawrence Erlbaum Associates.

Park, D. C., Nisbett, R., and Hedden, T. 1999. Aging, culture, and cognition. *Journals of Gerontology. Series B, Psychological Sciences and Social Sciences, 54*(2), 75–84.

Pavel, D. M., and National Center for Education Statistics. 1998. *American Indians and Alaska natives in postsecondary education*. Washington, DC: U.S. Department of Education, Office of Educational Research and Improvement.

Pennebaker, J. W., Chung, C. K., Ireland, M., Gonzales, A., and Booth, R. J. 2007. *The development and psychometric properties of LIWC2007*. Austin: LIWC. net.

Peshkin, A. 1997. *Places of memory: Whiteman's schools and Native American communities*. Mahwah, NJ: Lawrence Erlbaum Associates.

Peters, E., Burraston, B., and Mertz, C. 2004. An emotion-based model of stigma susceptibility: Appraisals, affective reactivity, and worldviews in the generation of a stigma response. *Risk Analysis, 24*(5), 1349–1367.

Philips, S. 1988. Similarities in North American Indian groups non-verbal behavior and their relation to early child development. In R. Darnell and M. K. Foster (Eds.), *Native North American interaction patterns*, 150–167. Quebec: National Museums of Canada.

Piaget, J. [1932] 1997. *The moral judgment of the child.* New York: Simon and Schuster.

Pierotti, R. 2011. *Indigenous knowledge, ecology, and evolutionary biology.* New York: Routledge.

Pierotti, R., and Wildcat, D. 2000. Traditional ecological knowledge: The third alternative [commentary]. *Ecological Applications, 10*(5), 1333–1340.

Pinxten, R., van Dooren, I., and Harvey, F. 1983. *Anthropology of space: Explorations into the natural philosophy and semantics of the Navajo.* Philadelphia: University of Pennsylvania Press.

Proffitt, J. B., Coley, J. D., and Medin, D. L. 2000. Expertise and category-based induction. *Journal of Experimental Psychology. Learning, Memory, and Cognition, 26*(4), 811–828.

Provenzo, E. F., and McCloskey, G. N. 1981. Catholic and federal Indian education in the late 19th century: Opposed colonial models. *Journal of American Indian Education, 21*(1), 10–18.

Quinn, N. 2000. Email exchange. In S. C. Strum and L. M. Fedigan (Eds.), *Primate encounters: Models of science, gender, and society*, 139. Chicago: University of Chicago Press.

Raina, D. 1997. Evolving perspectives on science and history: A chronicle of modern India's scientific enchantment and disenchantment. *Social Epistemology: A Journal of Knowledge, Culture and Policy, 11*(1), 3–24.

Reid, M. S. 2000. Toward effective technology education in New Zealand. *Journal of Technology Education, 11*(2).

Rogoff, B. 2003. *The cultural nature of human development.* New York: Oxford University Press.

Rosch, E., Mervis, C. B., Gray, W. D., Johnson, D. M., and Boyes-Braem, P. 1976. Basic objects in natural categories. *Cognitive Psychology, 8*(3), 382–439.

Rosebery, A., and Hudicourt-Barnes, J. 2006. Using diversity as a strength in the science classroom: The benefits of science talk. In R. Douglas (Ed.), *Linking Science and Literacy in the K–8 Classroom*, 305–320. Arlington, VA: National Science Teachers Association Press.

Ross, N. 2004. *Culture and cognition: Implications for theory and method.* Thousand Oaks, CA: Sage Publications.

Ross, N., Medin, D., Coley, J. D., and Atran, S. 2003. Cultural and experiential differences in the development of folkbiological induction. *Cognitive Development, 18*(1), 25–47.

Ross, N., Medin, D., and Cox, D. 2007. Epistemological models and culture conflict: Menominee and Euro-American hunters in Wisconsin. *Ethos* (Berkeley, Calif.), *35*(4), 478–515.

Rottenstreich, Y., and Hsee, C. K. 2001. Money, kisses, and electric shocks: On the affective psychology of risk. *Psychological Science, 12*(3), 185–190.

Rowell, T. E. 1966. Forest living baboons in Uganda. *Journal of Zoology, 149*(3), 344–364.

Rozin, P. 2007. Food and eating. In S. Kitayama and D. Cohen (Eds.), *Handbook of cultural psychology*, 391–416. New York: Guilford Press.

Rozin, P., Ashmore, M., and Markwith, M. 1996. Lay American conceptions of nutrition: Dose insensitivity, categorical thinking, contagion, and the monotonic mind. *Health Psychology, 15*(6), 438–447.

Rozin, P., Kurzer, N., and Cohen, A. B. 2002. Free associations to "food": The effects of gender, generation and culture. *Journal of Research in Personality, 36*(5), 419–441.

Rozin, P., Markwith, M., and Nemeroff, C. 1992. Magical contagion beliefs and fear of AIDS. *Journal of Applied Social Psychology, 22*(14), 1081–1092.

Rozin, P., Millman, L., and Nemeroff, C. 1986. Operation of the laws of sympathetic magic in disgust and other domains. *Journal of Personality and Social Psychology, 50*(4), 703–712.

Rozin, P., and Nemeroff, C. 2002. Sympathetic magical thinking: The contagion and similarity "heuristics". In T. Gilovich, D. Griffin, and D. Kahneman (Eds.), *Heuristics and biases: The psychology of intuitive judgment*, 201–216. New York: Cambridge University Press.

Ryan, R. M., and Deci, E. L. 2000. Self-determination theory and the facilitation of intrinsic motivation, social development, and well-being. *American Psychologist, 55*(1), 68–78.

Sachdeva, S. 2011. The norm of self-sacrifice. Ph.D. diss., Northwestern University.

Sadler, T. D., and Fowler, S. R. 2006. A threshold model of content knowledge transfer for socioscientific argumentation. *Science Education, 90*(6), 986–1004.

Scheman, N. 2003. Individualism and the objects of psychology. In S. Harding and M. Hintikka (Eds.), *Discovering reality: Feminist perspectives on epistemology, metaphysics, methodology and philosophy of science*, 2nd ed., 225–244. Dordrecht: Kluwer Academic Publishers.

Schroeder, C. M., Scott, T. P., Tolson, H., Huang, T. Y., and Lee, Y. H. 2007. A meta-analysis of national research: Effects of teaching strategies on student achievement in science in the United States. *Journal of Research in Science Teaching*, *44*(10), 1436–1460.

Schwartz, B. 1997. Psychology, idea technology, and ideology. *Psychological Science*, *8*(1), 21–27.

Schwartz, C. 1996. Political structuring of the institutions of science. In L. Nader (Ed.), *Naked science: Anthropological inquiry into boundaries, power, and knowledge*, 148–159. London: Routledge.

Scott, C. 1996. Science for the west, myth for the rest. In L. Nader (Ed.), *Naked science: Anthropological inquiry into boundaries, power, and knowledge*, 69–86. London: Routledge.

Seal, B. 1985. *The positive sciences of the ancient Hindus*. Delhi: Motilal Banarsidass.

Settee, P. 2000. *The issue of biodiversity, intellectual property rights, and indigenous rights*. Saskatoon: University of Saskatchewan Extension Press.

Sharpes, D. K. 1979. Federal education for the American Indian. *Journal of American Indian Education*, *19*(1), 19–22.

Shiva, V. 1993. *Monocultures of the mind: Perspectives on biodiversity and biotechnology*. London: Zed Books Ltd..

Shweder, R. A., Casagrande, J. B., Fiske, D. W., Greenstone, J. D., Heelas, P., Laboratory of Comparative Human Cognition, and Lancy, D. F. 1977. Likeness and likelihood in everyday thought: Magical thinking in judgments about personality [and comments and reply]. *Current Anthropology*, *18*(4), 637–658.

Shweder, R. A., Jensen, L. A., and Goldstein, W. M. 1995. Who sleeps by whom revisited: A method for extracting the moral goods implicit in practice. *New Directions for Child and Adolescent Development*, *67*, 21–39.

Simpson, A. G. B., and Roger, A. J. 2004. Protein phylogenies robustly resolve the deep-level relationships within Euglenozoa. *Molecular Phylogenetics and Evolution*, *30*(1), 201–212.

Sjöberg, L. 1998. Worry and risk perception. *Risk Analysis*, *18*(1), 85–93.

Slovic, P. 1999. Trust, emotion, sex, politics, and science: Surveying the risk-assessment battlefield. *Risk Analysis*, *19*(4), 689–701.

Smith, L. T. 1999. *Decolonizing methodologies: Research and indigenous peoples*. New York: St. Martin's Press.

Smith, L. T. 2006. Choosing the margins: The role of research in indigenous struggles for social justice. In N. K. Denzin and M. D. Giardina (Eds.), *Qualitative inquiry and the conservative challenge: Confronting methodological fundamentalism*, 151–174. Walnut Creek, CA: Left Coast Press.

Snarey, J. R. 1985. Cross-cultural universality of social-moral development: A critical review of Kohlbergian research. *Psychological Bulletin, 97*(2), 202–232.

Snibbe, A. C., and Markus, H. R. 2005. You can't always get what you want: Educational attainment, agency, and choice. *Journal of Personality and Social Psychology, 88*(4), 703.

Solomon, G. E. A., Johnson, S. C., Zaitchik, D., and Carey, S. 1996. Like father, like son: Young children's understanding of how and why offspring resemble their parents. *Child Development, 67*(1), 151–171.

Solovey, M. 2013. *Shaky foundations*. New Brunswick, NJ: Rutgers University Press.

Spelke, E. S. 1990. Principles of object perception. *Cognitive Science, 14*(1), 29–56.

Spelke, E. S., Breinlinger, K., Macomber, J., and Jacobson, K. 1992. Origins of knowledge. *Psychological Review, 99*(4), 605–632.

Spelke, E. S., Phillips, A., and Woodward, A. L. 1995. Infants' knowledge of object motion and human action. In D. Sperber, D. Premack, and A. J. Premack (Eds.), *Causal cognition: A multidisciplinary debate*, 44–78. New York: Oxford University Press.

Stahl, W. K. 1979. The US and Native American education: A survey of federal legislation. *Journal of American Indian Education, 18*(3), 28–32.

Steele, C. M., and Aronson, J. 1995. Stereotype threat and the intellectual test performance of African Americans. *Journal of Personality and Social Psychology, 69*(5), 797.

Stewart, G. M. 2010. *Good science? The growing gap between power and education*. Rotterdam: Sense Publishers.

Strike, K. A., and Posner, G. J. 1985. A conceptual change view of learning and understanding. In L. H. T. West and A. L. Pines (Eds.), *Cognitive structure and conceptual change*, 211–231. San Diego: Academic Press.

Strum, S. C., and Fedigan, L. M. (Eds.). 2000. *Primate encounters: Models of science, gender, and society*. Chicago: University of Chicago Press.

Sullivan, L., and Walters, A. 2011. Native foster care: Lost children, shattered families. *All Things Considered*. Retrieved from http://www.npr.org/2011/10/25/141672992/native-foster-care-lost-children-shattered-families on June 3, 2013.

Szasz, M. 1977. *Education and the American Indian: The road to self-determination since 1928*. Albuquerque: University of New Mexico Press.

Takasaki, H. 2000. Traditions of the Kyoto School of field primatology in Japan. In S. C. Strum and L. M. Fedigan (Eds.), *Primate encounters: models of science, gender, and society*, 151–164. Chicago: University of Chicago Press.

Teresi, D. 2002. *Lost discoveries*. New York: Simon and Schuster.

Tillman, L. C. 2009. Comments on Howe: The never-ending education science debate: I'm ready to move on. *Educational Researcher, 38*(6), 458–462. doi: 10.3102/0013189x09344346.

Tippeconnic, J. W. 1995. Editorial . . . on BIA education. *Journal of American Indian Education, 35*(1), 1–5.

Tippeconnic, J. 2000. Towards educational self-determination: The challenge for Indian control of Indian schools. *Native Americas: Hemispheric Journal of Indigenous Issues, 17*(4), 42–49.

Toulmin, S. E., Rieke, R. D., and Janik, A. 1984. *An introduction to reasoning*. New York: Macmillan.

Traweek, S. 1996. Unity, dyads, triads, quads, and complexity: Cultural choreographies of science. *Social Text, 46–47*, 129–139.

Trope, Y., and Liberman, N. 2000. Temporal construal and time-dependent changes in preference. *Journal of Personality and Social Psychology, 79*(6), 876.

Trope, Y., and Liberman, N. 2003. Temporal construal. *Psychological Review, 110*(3), 403–420.

Tsai, J. L., Louie, J. Y., Chen, E. E., and Uchida, Y. 2007. Learning what feelings to desire: Socialization of ideal affect through children's storybooks. *Personality and Social Psychology Bulletin, 33*(1), 17–30.

Tversky, A. 1977. Features of similarity. *Psychological Review, 84*(4), 327.

Tversky, A., and Kahneman, D. 1981. The framing of decisions and the psychology of choice. *Science, 211*(4481), 453.

Tversky, A., and Kahneman, D. 1983. Extensional versus intuitive reasoning: The conjunction fallacy in probability judgment. *Psychological Review, 90*(4), 293.

United Nations. 2007. Declaration on the Rights of Indigenous Peoples. Retrieved from http://social.un.org/index/IndigenousPeoples/DeclarationontheRightsofIndig enousPeoples.aspx on June 3, 2013.

Unsworth, S. J. 2008. The influence of culturally varying discourse practices on cognitive orientations toward nature. Ph.D. diss., Northwestern University.

Unsworth, S., Levin, W., Bang, M., Washinawatok, K., Waxman, S., and Medin, D. 2012. Young children learn comprehensive cultural frameworks of the biological world. *Child Development, 12*(1–2), 17–29.

Vaughan, E. 1993. Individual and cultural differences in adaptation to environmental risks. *American Psychologist, 48*(6), 673–680.

Vivieros de Castro, E. 2004. Exchanging perspectives. *Common Knowledge, 10*(3), 463–484.

Viveiros de Castro, E. 2010. In some sense. *Interdisciplinary Science Reviews, 35,* 318–333.

Ward, C., and Kennedy, A. 1994. Acculturation strategies, psychological adjustment, and sociocultural competence during cross-cultural transitions. *International Journal of Intercultural Relations, 18*(3), 329–343.

Warren, B., Ballenger, C., Ogonowski, M., Rosebery, A. S., and Hudicourt Barnes, J. 2001. Rethinking diversity in learning science: The logic of everyday sense making. *Journal of Research in Science Teaching, 38*(5), 529–552.

Warrior, R. A. 1994. *Tribal secrets: Recovering American Indian intellectual traditions.* Minneapolis: University of Minnesota Press.

Wason, P. C. 1960. On the failure to eliminate hypotheses in a conceptual task. *Quarterly Journal of Experimental Psychology, 12*(3), 129–140.

Waters, A. (Ed.). 2004. *American Indian thought: Philosophical essays.* Malden, MA: Blackwell.

Waxman, S., Medin, D., and Ross, N. 2007. Folkbiological reasoning from a cross-cultural developmental perspective: Early essentialist notions are shaped by cultural beliefs. *Developmental Psychology, 43*(2), 294–308.

Waytz, A., Cacioppo, J., and Epley, N. 2010. Who sees human? *Perspectives on Psychological Science, 5*(3), 219.

Waytz, A., Morewedge, C. K., Epley, N., Monteleone, G., Gao, J. H., and Cacioppo, J. T. 2010. Making sense by making sentient: Effectance motivation increases anthropomorphism. *Journal of Personality and Social Psychology, 99*(3), 410.

Weisberg, D. S., Keil, F. C., Goodstein, J., Rawson, E., and Gray, J. R. 2008. The seductive allure of neuroscience explanations. *Journal of Cognitive Neuroscience, 20*(3), 470–477.

Wellman, H. M., Cross, D., and Watson, J. 2003. Meta-analysis of theory-of-mind development: The truth about false belief. *Child Development, 72*(3), 655–684.

White, P. A. 1999. Toward a causal realist account of causal understanding. *American Journal of Psychology, 112*(4), 605–642.

Wilson, E. O. 1992. *The diversity of life.* Cambridge, MA: Harvard University Press.

Wilson, E. O. 1998. Integrated science and the coming century of the environment. *Science, 279*(5359), 2048–2049.

Winkler-Rhoades, N., Medin, D., Waxman, S. R., Woodring, J., and Ross, N. O. 2010. Naming the animals that come to mind: Effects of culture and experience on category fluency. *Journal of Cognition and Culture, 10, 1*(2), 205–220.

Wolff, P. 2007. Representing causation. *Journal of Experimental Psychology. General, 136*(1), 82–111.

Wolff, P., Medin, D. L., and Pankratz, C. 1999. Evolution and devolution of folkbiological knowledge. *Cognition, 73*(2), 177–204.

Wolpert, L., and Richards, A. 1997. *Passionate minds: The inner world of scientists.* New York: Oxford University Press.

Wortis, R. P. 1971. The acceptance of the concept of the maternal role by behavioral scientists: Its effects on women. *American Journal of Orthopsychiatry, 41*(5), 733–746.

Wu, S., and Keysar, B. 2007. The effect of culture on perspective taking. *Psychological Science, 18*(7), 600–606.

Yazzie-Mintz, E. 2007. Voices of students on engagement: A report on the 2006 High School Survey of Student Engagement. Center for Evaluation and Education Policy, Indiana University.

Yont, K. M., Snow, C. E., and Vernon-Feagans, L. 2003. The role of context in mother-child interactions: An analysis of communicative intents expressed during toy play and book reading with 12-month-olds. *Journal of Pragmatics, 35*(3), 435–454.

Zent, S. 1999. *The quandary of conserving ethnoecological knowledge: A Piaroa example.* Athens: University of Georgia Press.

Zuckerman, S. 1932. *The social life of monkeys and apes.* London: Kegan, Paul, Trench and Trubner.

Index